I0492404

Music and Medicine

Music and Medicine

Robert I. Levy, MD

Studies in Medicine, History and Culture

Volume 1

Series Editor: David B. Levy, PhD, MLS

2018

License Notes

This book is licensed for your personal enjoyment and information only. This book should not be re-sold to others. Educational institutions may reproduce, copy and distribute this book for non-commercial purposes without charge, provided appropriate citation of the source, in accordance with the Talmudic *dictum* of Rabbi Elazar in the name of Rabbi Hanina (*Megilah* 15a): "anyone who cites a teaching in the name of its author brings redemption to the world." This permission is intended for classroom use only. Small portions of the text may be posted on the Internet for review or study purposes.

Copyright David B. Levy

Version 1.0
תשע"ח
2018

ISBN: 978-1984930071

Sam
Sapozhnik
Publishers
Fiercely Independent.

To my parents

Robert I. Levy, MD
and
Ruth S. Levy, ע"ה

Studies in Medicine, History and Culture

Volume One

Music and Medicine

Volume Two

Essays in History of Nephrology

Volume Three

Essays in History of Medicine

Volume Four

Essays in the History of Allied Sciences

Volume Five

Pierre Rayer's *Tracté des Maladies des Reins et les Alterations de la Sécrétion Urinaire* A New Translation

Series Editor: David B. Levy, PhD, MLS

TABLE OF CONTENTS

Volume One: Medicine and Music

"If music be the food of love, O play on, It came over my ear like the breeze upon a bed of fresh violets"

Shakespeare, *As You Like It*

"If music be the food of love, play on, Give me excess of it; that surfeiting, The appetite may sicken, and so die."

Shakespeare, *Twelfth Night*

"There is geometry in the humming of strings, there is music in the spacing of heavenly spheres"

Pythagoras

"Music expresses that which cannot be said and on which it is impossible to be silent."

Victor Hugo

"Rhythm and Harmony find their way into the inward places of the soul."

Plato

"I know that the most joy in my life has come to me from my violin (and music)."

Albert Einstein

"Music is the universal language of mankind."

Henry Longfellow

"I call architecture frozen music."

Johann Goethe

א הַלְלוּ-יָהּ:
הַלְלוּ-אֵל בְּקָדְשׁוֹ; הַלְלוּהוּ, בִּרְקִיעַ
עֻזּוֹ.

1 Hallelujah.
Praise God in His sanctuary; praise Him in the firmament of His power.

ב הַלְלוּהוּ בִגְבוּרֹתָיו; הַלְלוּהוּ,
כְּרֹב גֻּדְלוֹ.

2 Praise Him for His mighty acts; praise Him according to His abundant greatness.

ג הַלְלוּהוּ, בְּתֵקַע שׁוֹפָר; הַלְלוּהוּ,
בְּנֵבֶל וְכִנּוֹר.

3 Praise Him with the blast of the horn; praise Him with the psaltery and harp.

ד הַלְלוּהוּ, בְּתֹף וּמָחוֹל; הַלְלוּהוּ,
בְּמִנִּים וְעֻגָב.

4 Praise Him with the timbrel and dance; praise Him with stringed instruments and the pipe.

ה הַלְלוּהוּ בְצִלְצְלֵי-שָׁמַע; הַלְלוּהוּ,
בְּצִלְצְלֵי תְרוּעָה.

5 Praise Him with the loud-sounding cymbals; praise Him with the clanging cymbals.

ו כֹּל הַנְּשָׁמָה, תְּהַלֵּל יָהּ:
הַלְלוּ-יָהּ.

6 Let every thing that hath breath praise the LORD. Hallelujah

Psalm 150

Introduction

by David B Levy, PhD

Remembrances of my Father From Childhood

I grew up as a youngster in a household filled with the love of listening to and playing classical music. My father was trained from an early age at Peabody music school. My father graduated from the Peabody preparatory school after an 8 year curriculum. After that for 2 years he studied under the guidance of Alexander Sklerevski in the Conservatory until my father was drafted at the age of 18 into the navy in World War II. He served on the aircraft carrier the US Saratoga.

But even on the ship, on the one day the aircraft carrier docked in port, my father went to the record store and bought sheet music. My father still remembers the piece: Chopin *Fantasie Impromptu* in C Sharp Minor, bought in San Francisco on 3/28/1945. He still plays this piece to this day 60 years later. The names Bach, Buxtehude, Rameau, Coperain, de Lully, Vivaldi, Handel, Mozart, Hayden, Beethoven, Brahms, Chopin, Schuman, Liszt, Debussy, Stravinsky, Shostakovich, and other muscicians were not just talked "about" but learned from.

My father plays their music. My father can hear played in the mind just looking at their musical scores. My father also reads a great amount of texts in musicology and music. Books by Charles Rosen such as *the Baroque Style, the Classical Style, the Romantic Style* were texts the whole family would read together. Further works such as *Emblems of the Mind- the Inner Life of Music and Mathematics* by Edward Rothstein were discussed around the table. With regards to the relationship of music and science my father loves Oliver Sacks book: *Musicophilia: tales of Music and the Brain*. We both heard Dr Sacks speak numerous times at the series of monthly lectures hosted by Johns Hopkins Medical School on Medicine and the Humanities that was organized by Drs. Richard Macksey and Neurosurgeon George Udvarhelyi.

Both scholars Macksey and Udvarhelyi valued and put a premium on well rounded physicians versed not only in science but the Humanities. Udvarhelyi worked clandestinely in Nazi-occupied Budapest with Swedish diplomat Raoul Wallenberg to save Hungarian Jews from the Holocaust. He with the Homewood campus' humanities professor Richard Macksey, founded the Hopkins medical school's Office of Cultural Affairs in 1977. It was the first of its kind in an American medical school and brought to the East Baltimore medical campus prominent writers, actors, musicians, artists and authors as Isaac Bashevis Singer, Aaron Copland, Richard Leakey, and others for lectures and musical performances. Macksey and Udvarhelyi planned this excellent medical humanities curriculum. I would attend these lectures and concerts with my father at the Johns Hopkins Medical Library.

My father admires the altruistic Dr. Albert Schweitzer. My father's copy of the 2 volume set on Bach by Schweitzer was required reading in the Levy household etc. Schweitzer gave up a prominent medical career in America, being a graduate of Harvard, and instead served underserved communities in Africa.

To one cellist who played duets with my father, my father gave to them as a gift an expensive facsimile of the works of Beethoven as my father loves good books, as did the Hopkins physician William Osler. In the volume on the history of medicine there are a number of essays on Osler. Osler's library is gifted to the University of Montreal, in which my father did research. My father wrote an essay about Osler's love of books, noting that his favorite book was *Religio Medici* (*The Religion of a Doctor*) by Sir Thomas Browne. Osler collected many different editions of *Religio Medici*, and Osler sought to gather them in his private library.

My father loves researching etymologies in Greek, Latin, German, French, and Italian especially as they help to better appreciate music. My father had me start learning German to appreciate music, by not only reading the poetry of Heinrich Heine, Maria Rilke, and the work of Franz Kafka, but also listening to German Lieder. Dietrich Fischer-Dieskau's recordings, best known as a singer of Franz Schubert's Lieder, particularly "Winterreise" with accompanist Gerald Moore and Jörg Demus were avidly listened to as were *Die schöne Müllerin*. Johann Wolfgang von Goethe (1749-1832), wrote numerous poems, from which Franz Schubert made songs from the lyrics of his poems. Goethe's poems were also set to music by Robert Schuman, Gustav Mahler, Hector Berlioz, and Frank Liszt. In College at Middlebury language school my father would send me recordings of Schubert with the text of Goethe to better learn German. For instance the *Erlkönig* is Goethe's most famous balled. It depicts the death of a child assailed by a supernatural being. The haunting piece describing a father's lament over his son's dying moments as the father carries the son through the forest wilderness The young boy is being carried at night by his father on horseback. My father would send me the recording form, with accompanying German written text in order to learn a language essential to any musical education.

When my father went to an opera he did not just go to be merely entertained. Rather if he saw Verdi's *Falstaff* he did extensive research on seeing how Shakespeare's 5 plays in which Falstaff appears, including the play *Henry IV* Part I and Part II, compared in text with Verdi's score, discovering important differences. My father also studied the difference between Shakesepare's *Macbeth* and Verdi's opera by reading works such as the sourcebook edited by David Rosen and Andrew Porter published by W.W. Norton & co. Similarly the opera *Othello* was compared to Shakespeare's tragedy. My father took the whole family to the Charles Movie theater to see Placido Domingo in Zeffirelli's film *Otello*. Discussion around the table involved how Shakespeare' play was reinterpreted by Verdi's opera. To this day my father relishes watching at the movie theater Operas on film. My father is overcome with emotion when listening to Verdi's 1841 opera Nebucco when the Hebrew slave chorus begins their chant. I actually attended a lecture on Verdi's Nebucco with my father on how Italian Nationalism in this opera operated in tandem with the rise of Jewish nationalism.

Other musicians inspired by Shakespeare's work include Benjamin Britten, Sibelius, Tchaikovsky, Liszt, and Mendelssohn. Diverse composers such as Henry Purcell, Carl Orff, Ralph Vaughan Williams, and Benjamin Britten wrote musical comedies based on Shakespeare's *A Midsummer Night's Dream*. My father is still a season ticket holder, member of the Folger Shakespeare Theater in Washington DC and visits their library to research many topics including Renaissance music and allusion to music in Shakespeare's plays.

If my father saw the opera *The Marriage of Figaro* he made it a point to spend hours studying the play. The Marriage of Figaro (French: *La Folle Journée, ou Le Mariage de Figaro*) is a comedy in five acts, written in 1778 by Pierre Beaumarchais. Likewise, if my father went to a performance of the work *Pelléas et Mélisande*, Op. 80, a suite derived from music by Gabriel Faure, inspired from Maurice Maeterlinck's play of the same name, my father studied the similarities and differences of the score and drama.

This kind of comparison between musical score and the subtext of the drama was also investigated for example in the following works:

1) Musical treatments by Telemann, Mendelssohn & Massenet of Don Quixote sent my Father to study the original text by Cervantes.

2) Rossini's Cinderella and ballet by Prokofiev and Sleeping Beauty ballet by Tchaikovsky and Hansel & Gretel opera by Humperdinck sent my Father to read the original Fairy Tales.

3) Likewise if he researched Rimsky-Korsakov's Scheherezade he read the text of the Arabian Nights.

4) Benjamin Bortleau's Paul Bunyan sent my Father to read the legendary story.

5) The opera Carmen by George Bizet was inspired by a true story by Mercurée.

6) The novel War and Peace by Tolstoy also served as the source for an opera by Prokofiev and Tolstoy's *Anna Karenina* was realized in opera by Scottish composer Iain Hamilton.

7) If my father heard Offenbach's Dafoe he also read the novel Robinson Crusoe by Daniel Defoe.

One vivid memory I have is traveling to NY Met with my father on a Sunday to witness a performance of Schoenberg's opera *Moses and Aaron*. My father taught me to see how Schoenberg referenced the Midrash in attributing to Moses *spreche stimme*, or stuttering like music (staccato), while Aaron as the gifted speaker is given virtuosic Italian arias to sing in fluid song. At the calling (*Gattung*) of Moshe at the burning bush, out of humility Moshe who was *anuv meod* (very humble), and not just humble, as the Rambam notes, one should go to the extreme in this virtue of humility in *Hilchot Deot*. Anyone who meets my Father notes that he is very humble, unassuming and modest. Other prophets also out of humility sought to flee from their prophetic calling such as Jeremiah who claims he is just a lad (naar) and Isaiah who says he is of unclean lips (see *Haftorah Yetro*). Moses refers to himself as "heavy of mouth and heavy of tongue" so G-d tells Moshe that Aaron who is fluent in speech will be his spokesperson. We thus learned from the Schoenberg opera how did Moshe become a stutter? This is based on the Midrash that Moshe was a stutterer because he reached for a burning coal that touched his lips, rather than a pot of gold when shown both. This was itself prompted by Pharoah who once witnessed the baby Moshe raised in the Palace who reached for the Pharoah's jeweled gold crown. Pharoah being afraid of losing his throne, asked his astrologers and counselors the meaning of this action of the infant. One of the king's counselors, however, suggested that they should first test the boy and see whether his action was prompted by intelligence, or he was merely grasping for sparkling things. Thus two bowls were set before young Moses. One contained gold and jewels, and the other held glowing coals. Moses reached out for the gold, but an angel directed his hand to the coals. Perhaps the Torah is teaching that thought should precede rhetoric, as Moshe is the chief ot the prophets (לא קם בישראל כמשה עוד, נביא ומביט את תמונתו). According to Rambam, Moshe's *nevuah* was supreme because it was based on reason while the other prophets employed creative metaphors via the imagination. Moshe made his imagination ministerial to his reason and his revelation was in the form of law. Imagination is the mirror that shines less clearly than reason—אספקלריא לא מאירי.

More recent operas my father took me to included *Porgy and Bess*, with the song *Old man River* which I loved with much of the music based on George Gershwin's work.

Besides Opera, my father also enjoys Cantorial music that often employs operatic techniques and methods. I heard with my father Chanuaka concerts hosted by Beth Tefilah congregation in Baltimore, where greats such as Yakov Motzin, Avraham Albrecht, and Dudo Fisher sang a diverse repertoire. In an earlier generation my father followed the careers of Cantors turned opera singers such as Richard Tucker, Jan Pearce, Hugo Weisgal, and Yoselle Rosenblatt. Rosenblatt was a descendent of the *mikubal* (Kabbalist) Rabbi Astropolar from Eastern Europe. Rosenblatt was the son of a Cantor from Kiev. The autobiography by Rosenblatt describes how in eastern Europe from Poland on westward, young *precentors* traveled about from congregation to congregation, bringing new melodies. These wandering minstrels, of their craft, often brought a vocal orchestra but sometimes a choir, designated "*meshorerim*" or song-makers. *Meshorerim* acted as prompters for cantors and in some instances served as repositories of melodies. Some communities, such as Prosnitz (Moravia), embraced them enthusiastically, while others condemned the practice. While orthodox shuls prohibited use of the organ on Shabbos based on Halakhah, reputedly, the organ built for the Dohány Street Synagogue in Budapest in 1859 was sufficiently powerful and renowned to have been played by both Franz Liszt and Camille Saint-Saëns! The function *meshorerim* was likened to that of the youthful Levites who had stood below the platform of the singers in the Temple on the steps, to "give flavor to the song" ('Ar. 13b; comp. Yoma 38a). This is the tradition out of which Moshe Oysher (1907-58), Gershon Sirota (1874) 1943), Moshe Koosevitsky (1899-1966), David Roitman (1884-1943), Berele Cleagry (1892-1940), Mordechai Herschman (1888-1940), Pinchas Segal (1900-1971) and the famous Rovno cantor" Mordkhe Shvartsenberg emerged.

Admirably Rosenblatt would not perform on Shabbos and so his opera career was a *kiddush Hashem*. Notably he passed on, on a film shoot in Eretz Yisrael for a singing part. It is the voice of Rosenblatt that is found in the early version of the film *the Jazz Singer*. Also my father heard in person Hugo Weisgal many times at Chizuk Amuno Congregation.

My father would buy CD recordings and distribute them to his patients and friends as a physician in private practice, hoping the patients would not only enjoy the music, but be therapeutically helped to be healed by it. Rambam wrote a *Responsa* on listening to music (edited by President of the Baltimore Hebrew University, Israel Efros). The Rambam notes music's therapeutic value, as discussed in three works:

1) I Goldziher, "Das Gutachten des Maimonides ueber Gesang und Musik" in Monatsschriften fuer Geschichte und Wissenschaft des Judentums vol 22 (1872), p174-180

2) Schirmann, Haim, "Ha-Rambam ve-ha-Shira ha-ivrit" in Moznayim vol 3 (1935) p. 434-435 * notez Bien, Schirmann was a scholar of the Cairo Geniza poetry and a very accomplished violinist and when his apartment in Tel Aviv caught on fire he ran into the fiery blaze to save his violin at great risk to his life, his one prized possession!

3) I Adler, "La pratique musicale savant dans quelques communites juive en Europe aux XVIIe-Xviiie siecles, Paris: La Haye, 2 vols, 1966

For one of my father's birthdays, he bought out a performance of an all Mozart Concert seating hundreds of his friends and colleagues on a motzi Shabbos at Peabody Music Conservatory, some who had never been exposed to classical music before. If this was not just therapeutic it exposed many to classical music for the first time. This concert was conducted by the musician Edward Polochick, now Artistic Director of *Concert Artists of Baltimore*. My father also donated rows of tickets, so that music students who could not afford Meyerhoff symphony concerts, could attend the performances.

My Father's Childhood

Perhaps my father's love of music which grew throughout the years was instilled by his parents and grandparents. His grandmother Linnie Eisenberg was an accomplished pianist, and my father still remembers to this day Linnie playing Franz Liszt's Hungarian Rhapsodies, with brother Abraham Eisenberg. Linnie and Abe were the children of Gustav and Esther Eisenberg from Hungary. Grandmother Linnie Eisenberg was an accomplished pianist who married Moses Bear. Moshe Bear owned a clothing store in Lonaconing Maryland. Linnie was the daughter of Gustav and Esther Eisenberg from Hungary. Linnie's husband was Moses Bear who was the son of Hartz Bear from Bavaria and Barbetta (Bracha) Jandorf Bear from Wurtenberg Germany.

As a young child my father was taken to the Abraham Eisenberg house on Park Heights Avenue, to play the piano. My father's parents, Dr. Charles Levy and Ruth Bear Levy, managed to scrape together funds to buy my father a relatively inexpensive Kindal piano. To this day my father plays upon this same piano, refusing to upgrade to a Yamaha or Steinway, as the piano gifted by his parents provides a kesher (link) to memories of his blessed parents (ztsl), who encouraged my father to play piano and instilled a love for classical music.

In my father's adulthood, on Sundays, fellow physician European trained colleague friends such as Drs. George Vash who played the violin, opera singer Sylvia Stuck, countless Professors (such as the Roses), and students from Peabody Music Conservatory, and the cellist from the Washington Symphony, and many others visited the house to make music in the living room. That truly was a living room alive with the hum and buzz of stringed instruments at play. Sometimes musicals took place whereby musicians gave a 15 minute spiel about the piece they would play, and then everyone performed their musical offerings in the living room.

There was one remarkable musician named Spencer Hoffman, a Mozart expert, who served as a Musician for the Naval Academy, while his true love was classical music. My father made sure I and my two sisters took piano lessons from Hoffman, who was against "public performances." Hoffman believed that classical music was best appreciated in these small musical get togethers by knowledgeable amateurs and experts who loved music, rather than those who went to concerts "to be seen and socialize". Hoffman taught music to a few select students. My father managed to convince him to try to teach me although I did not show the promise of my father's expertise and disciplined devotion in sacrificing countless hours to playing and practicing. Hoffman would record the accompanying instruments to the piano. For instance the first piece my father and Hoffman had me try to learn, after becoming proficient in musical notes by the mnemonic (EVERY GOOD BOY DOES FINE/and FACE), and learning to count musical 'time' was Beethoven's 9th symphony, with Hoffman recording the other musical instruments while I chimed in on the piano. Of course my teachers directed me to the German text of Schiller's *Ode to Joy* that forms much of the lyrics in the fourth movement of Beethoven's symphony. Hoffman smoked delicious pipe tobacco, cherry, vanilla, and other blends that he filled the room with aromatic smoke like incense, while off it seemed in a trance while playing music.

While Hoffman was not gung ho about attending public performances, my father loves to do so and can recall many concerts he heard such as hearing Toscanini, Pinchas Zuckerman of the Israeli Philharmonic, and the Israeli Dalia Atlas virtuosically playing the piano. Atlas who was born in Haifa Israel, graduated from the Music Academy of Jerusalem. My father also relishes concerts by Gil Shaham whose family returned to Jerusalem when Gil was two. At the age of seven, Shaham began taking violin lessons from Samuel Bernstein at the Rubin Academy of Music in Jerusalem.

Recently my father and I heard a trio of 3 Israeli musicians name (a) clarinetist Alexander Fiterstein, (b) pianist Alon Goldstein, and (c) cellist Amit Peled played a piece titled "Hasidic Fantasy" by Israeli cellist and composer Joachim Stutschewsky (1891-1982) who was a European classically trained musician who fled Europe to Israel during the Holocaust. Joachim was an Ukraine-born Austrian and Israeli cellist, composer, musicologist. His father, Kalmen-Leyb Stutschewsky was a clarinetist. He was married twice. His first wife was the Swiss cellist Rewekka "Wecki" Schein. Stutschewsky's archive can be found in the Felicja

Blumental music library in Tel Aviv at הספרייה במרכז פליציה
בלומנטל למוזיקה

My Father felt the best concerts he went to in the 1940s were when he heard Rachmaninoff play two times. My Father's mother encouraged my Father as a youngster to practice the C Sharp Minor Prelude of Rachmaninoff "just in case Rachmaninoff would ask my Father as a kid to play this piece."

Although my father takes pride in the high culture of Israeli musical life, he does believe that good music is a universal language that speaks to all persons regardless of culture, geographical, socio-economic, and other differences. My father avidly attends other concert series such as the *Candle Light Artists of Baltimore* who give concerts on Saturday nights and the *Schriver Hall Concert Series*. When my father travelled outside of Maryland (which was rare due to his extremely busy schedule) he always sought out concerts. For example when I moved to NY I remember going with my father to the 92nd St Jewish Y to hear the *Beaux Arts Trio, Emerson String Quartet*, and *Ying Yang Quartet*, and other classical music ensembles. When he heard Beethoven's *Four Quartets* my Father was eager to read T.S. Eliot's poems of the same name.

Despite Hoffman's criticism of musical public performances, my father loves going to concerts, but not for mere entertainment. Rather before the event, he studies the score/sheet music preparing himself to better understand and appreciate any "interpretation" of a musical work. Perhaps Hoffman might be analogized to the GRA (Vilna Gaon) although one in the field of music and the GRA a Talmud scholar. The analogy is that while the GRA who davoned with a select few in a Kloyz or in a private minyan shunning davoning in a large synagogue, that was next door in Vilna. In Vilna at the time there were a number of small minyaneim such as the *Gmilut Hesed Kloyz*, the Painter's *Kloyz*, the Artizan's *Kloyz Chevra Po'alim*, the *Shtibl* of the Koidanover Hasidim, the *Shtibl* of the Lubavich Hasidim (later *Tiferet Bahurim Kloyz*),

On one occasion I drove with my father to some remote location, to a cabin in the woods where we enjoyed a private get together concert where the pianist Richard Goode spoke and played in a very small group. Goode is known best especially for his interpretations of Mozart and Beethoven and his recordings of the Goldberg variations by Johann Sebastian Bach, French Suites, and English Suites which are truly sublime (das Erhaben).

Medicine and Healing

To this day my father loves to play music and the soul takes flights and journeys on winged notes of a musical odyssey, as my father appears to almost go into higher realms, as he really resonates with the piece he may be playing, as if in more spiritual worlds, far far away in heaven's distant gardens (*die himmelferne garten*). In my father's small apartment today, a key place on a music stand is a painting of Brahms, sitting at the piano smoking a pipe almost in the same pose of being transported into higher worlds while playing music.

While medicine strives to heal the physical body music heals the spiritual body, although any holistic well rounded physician, if they are a good one, will take to heart the oath that the physician, dedicated not only to promote longevity of life, but quality of life, and this means appreciating and learning from Music. In Rambam's *Responsa* on listening to music, the Rambam, as well as throughout his medical writings, as Fred Rosner shows, refers to the psycho-somatic aspect of music. Music can sometimes reduce pain and take one out of suffering, to transcend this world, but return with greater appreciation for and love for this world and love for human beings. In one musical analogy G-d created this world as if writing a musical composition of the symphony of life. In Franz Shubert's lieder *Auf dem Wasser zu singen*, which takes place with someone in a row boat on glacial lake in the Alps looking up at the stars, while gliding along, the sound of a trout jumping out of the water and making a splash is a kind of "amen" to the beautiful surroundings. Thus when Franz Rosenzweig at the end of his life suffered from Lou Gehrig's disease, the one comfort the philosopher experienced was listening to phonograph records of classical music. My father and I attended a lecture of Dr. Barbara Galli on this very subject of Rosenzweig's love of classical music in a series on German Jewish culture lectures which featured other scholars such as Sander Gilman, Leon Botstein, Mendes-Flohr and others.

Musical Vacations and Music Mission

My father would travel with my mother in the summer during his precious little time off from work, to Aspen Music festival, Bard College music seminars, Santa Fe Opera fest, and the Tanglewood festival. At Tanglewood for instance my father heard works conducted and written by Leonard Bernstein, such as the Jeremiah symphony, Chichester Psalms, Four Sabras, Haskivenu, Israelite Chorus, and Song of Songs. Bernstein's *The Unanswered Question*, was read in our house amongst other classics in music. When my Father researched Bernstein's 1956 Operetta *Candide* he studied Voltaire's works on the subject. My father, when he heard Bernstein's *West Side Story*, researched its bases on Shakespeare's *Romeo and Juliet*. Likewise Bernstein's Symphony No. 2 sent my

Father to explore poetry with Auden's *The Age of Anxiety* on which Bernstein's Symphony is based.

From Pythagoras and Kepler and the music of the spheres, to Einstein and his violin, my father made me appreciate, from an early age that wherever we turn, music and mathematics, science, and art, bear a strong relationship when we consider that for millennia medicine itself was considered a medical art, at least from Galen on.

My father loves classical music of all sorts, as an insider who has studied musicology, and plays the piano. A particular favorite is Haydn, who tagged to his musical scores "a praise of love for G-d," *Laus Deo*, a phrase penned at the end of his scores, as well as incorporated in his song, "*Aus dem Danklied zu Gott*" (From the song of Thanks to God). In the "*Abenlied zu Gott*" (Evening song to God) Haydn adopted the poetry and spirit of genuine humility of the devout Protestant Christian, Fuerchtegot Gellert, "*Her der du mir das Leben/ Bis diesen Tag gegeben/ Dich bet ich kindlich an/ Ibn bin viel zu geringe/ Der Treue, die ich singe/ Und die du heut an mir getan* [L-rd you who have given me life up to this very day, to you I pray like a child/ I am much too unworthy/ of the faithfulness of which I sing, / and which you show to me today].

My father talks with excitement when noting that while Handel's oratorios clearly reveal a religious biblical consciousness as in: Esther, Samson, the resurrection, Deborah, Joseph and his brothers, Judas Macabaeus, Joshua, Solomon, Belshazzar, Jeptha, etc. Haydn's oratorios such as the Creation (*Die Schopfung*) sings of the omnipotence, majesty, and infinite wisdom of G-d. It opens, "*Im Angange schuf Gott Himmel und Erde* (in the beginning God created the Heaven and the Earth). Haydn's work continues: *Mit Staunen sieht das Wunderwerk* (The marv'lous work beholds amazed) with the crescendo of the creation account whereby the chorus sings, "*Stimmt an die Saiten*" (awake the harp) whereupon the tenor Uriel, sings in recitative, "*Und Gott Sprach: Es sei'n Lichter an der Feste des Himmels*" (Let there by lights in the firmament of the heaven)- whereby in Haydn's symphony drawing on a verse from Psalms: Die Himmel erzaehlen die Ehre Gottes (The heavens declare the glory of G-d, *Hashamayim misaprim ha-kavod Kail*). Although Haydn's religious orientation is Christian, my father recognizes a deep religious sentiment emblematic of a deep faith. However as my father has pointed out to me, this is problematized by the reception history of Haydn's work: *The Apothecary* that has anti-semitic images of Jews.

My father also enjoys celebrating the philo-semitism of composers like Dmitry Shostakovich who incorporated Yiddish folk song melodies into a number of his pieces and devoted one musical offering to a Jewish friend in a concentration camp.

In Joachim Braun's "The Double Meaning of Jewish Elements in Dmitri Shostakovich's Music," *The Musical Quarterly* 71:1 (1985) 68-80, the musicologist divides Shostakovich's Jewish works into three periods. The first period were works composed between 1943 and 1944. While Braun includes the last movement of the Second Piano Trio and Shostakovich's complete orchestration of Venyamin Fleishman's opera *Rothschild's Violin* which was the original cause of his interest in Jewish music. During this period Shostakovich's reasons for including Jewish elements in his works seems to have been principally aestheteic although Shostakovitch was always repelled by anti-Semitism, and wished to express his opposition to the growing anti-Jewish feeling within the USSR. Thus Shostakovitch incorporated Yevgeny Yevtushenko's poems denouncing anti-Semitism into his 13th Symphony (*in B-*

flat minor (O. 113)), which commemorates Babi Yar, *Babiy Yar)*, a site of 1941 massacres carried out by German forces and by local Ukrainian collaborators during their campaign against the Soviet Union in World War II. This Babi Yar syphony was completed on July 20, 1962 and first performed on 18 December 1962 by the Moscow Philharmonic Orchestra. In the Babi Yar movement, Shostakovich and Yevtushenko transform the mass murder by Nazis of Jews at Babi Yar, near Kiev, into a denunciation of all anti-Semitism in all its forms. Shostakovich sets the poem as a series of episodes—the Dreyfus affair, the Białstok pogrom, and the story of Anne Frank.

My father takes pride to this day in the fact that many accomplished classical musicians are Jewish. This is the opposite of the hateful *Lexikon der Juden in der Musik* which was a Nazi sponsored encyclopedia first published in Germany in 1940, which black listed individuals involved in the music industry who were defined under Nazi Racial Laws, as 'Jewish' or 'half-Jewish' (*halb Juden, mischlingen*). Such an anti-semitic work sought to discredit the great Jewish musicians of the past such as Goldmark and Offenbach (sons of Chazanim), Fromental HaLevy (author of *La Juive*), Felix Mendelsohn, Berthold Goldschmidt. As well as contemporary artists like Ernst Bloch (musical themes), Atonal scale of Schoenberg, Erich Korngold, Aaron Copeland, Fritz Kreisler, Gustav Mahler, Arthur Schnabel, Alfred Shnitke, Kurt Weill, Leonard Bernstein, André Previn, Anton Rubinstein, Hafetz, etc. The Nazis blacklisted these musicians as well.

A note on the *avant garde* classical music of Philip Glass-his mother was the head librarian at the high school my Father attended.

My father also appreciates Biblical themed music. Italian Renaissance Rabbis Shlomo de Rossi, Judah ben Yosef

Moscato (1530-1593), Leon Modena (1571-1648), and Dr. David Abraham ben David Portleone (1542-1612) who set psalms and liturgical passages in Hebrew to music were areas of great love as were the works of Handel's oratorios on Biblical themes. My father read the work Israeli musicologists such as those by Don Haran, Joachim Braun, and other academics.

My father also shows great interest in emigre German Jewish musical culture (i.e. (1) music made by Jews,(2) music in Jewish style, (3) music with Jewish content, etc.) that was largely displaced to either the US or Israel in the 5th or German Aliyah, represented by musicians such as Paul Ben-Haim, Haim Alexander, Tzvi Avni, Paul Ben-Haim, Yehezkel Braun, Abel Ehrlich, Robert Lachmann, Ben-Zion Orgad, Erich Walter Sternberg, Josef Tal , that drew from Central Europe to build a modern musical culture after 1948, as the result of Nazi persecutions between 1933-1945.

This transnational dislodgment of German Jewish émigré musicians also fractured in resettlement in America as represented by the works of Nazi discriminated against musicians such as Gustav Mahler, Arnold Schoenberg, Kurt Weill, etc. My father continues to appreciate the contributions of this Jewish émigré musical culture (Bildung) after the Holocaust, conceived by some as a Hegelian "Stunde nulle" (zero hour) in history and the Beginning of the End (*Der Anfang des Ende*). The piece *Quatuor pour la fin du temps, (Quartet for the End of Time)* by French composer Olivier Messiaen can be viewed as another piece echoing 'end of time' that the Holocaust constitutes in Hegelian aufhebung. Messiaen was 31 years old when France entered World War II. He was captured by the German army in June 1940 and imprisoned in a prisoner-of-war camp in Gorlitz, Germany. While in transit to the camp, Messiaen showed the clarinetist Henri Aloka, also a prisoner, the sketches for what would become *Abîme des oiseaux* (*Abyss of the Birds*). Two other professional musicians, violinist Jean le Boulaire and cellist Etienne Pasquie, were among his fellow prisoners. Messiaen wrote a short trio for them; this piece developed into the Quatuor for the same trio with himself at the piano. The quartet was premiered at the camp, outdoors and in the rain, on 15 January 1941.

Yet despite the tribulations of the unique persecutions these Jewish émigré musicians suffered from, they continued historically traditions from past European Jewish vibrant musical expressions. A classic photo on the founding of Hebrew University in Jerusalem, with Rav Kook on the stage, next to the Chief Sephardic Rabbi, and academic Judah Magnes, and other Professors with Lord Balfour in the center giving his blessings to the founding of Hebrew University in Jerusalem, is further noteworthy that some of the greatest musicians from Europe, a number who survived the Nazi Holocaust, are in the foreground with musical instruments playing at the initial inaugural commemoration of the University that would employ some of the finest European academics transplanted to Jerusalem such as Martin Buber, Ernst Simon, Gershom Scholem, and Chaim Weizman.

Historical Context of Ancient and Medieval Music

In my path in Jewish studies my father encouraged me to take a course specifically on the ten biblical songs in the Tanakh, namely (1) Song of Adam Rishon (Psalm 92); (2) Shirat ha-Yam (Az Yashir Moshe, Shmos 15); (3) Az Yashir Yisrael (Bamidbar 21:17-20); (4) HaZinu (Devarim 32:1-43); (5) Yehoshua 10:12-14; (6) Devorah in Shoftim 5; (7) Chana in (Shmuel Alef 2:1-10); (8) Dovid HaMelekh (II Shmuel

22:1-51); (9) Shir HaShirim shel Shlomo, (10) Yeshaya 26:1-10, perhaps with hopes that this would help train my own appreciation of music and enhance a kindred musical *neshama*. Also there was encouragement to study however briefly with a cantor and gifted baal korei who also played the baroque flute, Michael Lesley.

I believe that the soul spark of my father's love for music, however is not all from more recent classical music, but stems back to the ancient origins in Jewish history.

In the times of the *Beit HaMikdash* music surrounded the *korbanot*. Amos [6:5] and Isaiah [5:12] show that the *seudot* following *korbanot* often were attended with music, and from Amos [5: 23] we learn that songs become a part of the regular Temple liturgy. Victorious military generals were welcomed with music on their homecoming (Judges 11: 34; I Sam. 18: 6), and music accompanied the dances at harvest festivals (Judges 9: 27, 21: 21) and at the accession of kings or their marriages (I Kings 1:40; Ps. 14: 9). Shepherds cheered their loneliness with their reed-pipes [I Sam. 16: 18], and Lam. [5: 14] shows that youths coming together at the gates played stringed instruments.

In my path of Jewish studies for instance when I was learning *Maseket Rosh Hashanah* with a rabbi my father made sure I asked the Talmud teacher about the musical significance of the Shofar. Near the Temple Mount in Jerusalem, archeologists have excavated an inscription stating "blow the shofar on Rosh Hodesh."

There are a total of 100 Shofar blasts according to the Talmudic tractate with many esoteric meanings and correlations. My father intuitively seemed very interested in knowing why some blasts are broken resembling cries of a human being, some are elongated. The Tekiah is 1 *blast*, the Shevarim is 3 wavering continuous *blasts*, and the

Teruah is 9 staccato *blasts*. My father rightly identified intuitively that there was some meaning and significance very important behind these types of blasts. In addition to Rambam's explanation that the Shofar wakes one up from their spiritual slumbers, or Rav Saadia Gaon's 13 reasons why a shofar is blown found in Agnon's *Sefer HaYamim Noraim*, my father's question was able to penetrate to the mystical reasons for the intrinsic character of the blasts that signify something more profound than can be conveyed by language. *Lieder ohne Wort!*

When I brought home a rare transcription of *Megillat Esther* trope around Purim that I accidentally found in the stacks of a library, my father was delighted by the challenge of playing music backwards as the music flowed in the direction of Hebrew right to left, rather than as English left to right! Although my reading of music is very primitive I often sat as a boy on the piano bench with my father and attempted to at least be a page turner. The quick nod of my father's head and blink signaled me to turn the page.

Shalom Carmy has written a wonderful essay in Tradition in 2014 *"Music of the left hand : personal notes on the place of liberal arts education in the teaching of R. Aharon Lichtenstein."* Indeed music should play a most important part in the Rabbinic curriculum.

It is my father's musicality that inspired my own interest in Jewish music, of which the following expanded excerpt adorns the library guide on the Jewish Arts for which I provided content where I work:

While the fine arts are in essence corporeal, music as an art is purely of the spirit and the most spiritual of all the arts. The work, *"Ta'amei HaTa'amim"* (The reason for the trope), authored by the Masorite Asher ben Asher who lived in Tiberia sometime in the 7th century likens the art of cantillation to giving "soul" to the letters of the text. In this work when the *baal korei* chants the text correctly according to musical trope this act of music "opens the gates in Shamayim." While the letters of the text are the body of the torah, the music is the soul according to the Masorite Rabbi Asher ben Asher.

Music and the Bible

Music in antiquity was employed by many prophets in order to prophesy and go into ecstatic states as noted by Dr. Moshe Idel. I was briefly in a course with Idel on Rabbi Abraham Abulafia where I learned about Rabbi Abulafia's interest in music to thrust the rabbi into trances of prophetic inspiration. The source of music as prophetic inspiration goes back to the *Tanakh*. The ecstasy of the Prophets was stimulated by dancing and music (I Sam. 5 5, 10; 19: 20).

In sefer Shmuel a band of neveim with psaltery, timbrels, pipes, and harp are said to be sprung into prophecy by music

ה אַחַר כֵּן, תָּבוֹא גִּבְעַת הָאֱלֹהִים, אֲשֶׁר-שָׁם, נְצִבֵי פְלִשְׁתִּים; וִיהִי כְבֹאֲךָ שָׁם הָעִיר, וּפָגַעְתָּ חֶבֶל נְבִאִים יֹרְדִים מֵהַבָּמָה, וְלִפְנֵיהֶם נֵבֶל וְתֹף וְחָלִיל וְכִנּוֹר, וְהֵמָּה מִתְנַבְּאִים.

5 After that thou shalt come to the hill of God, where is the garrison of the Philistines; and it shall come to pass, when thou art come thither to the city, that thou shalt meet a band of prophets coming down from the high place with a psaltery, and a timbrel, and a pipe, and a harp, before them; and they will be prophesying.

י וְצָלְחָה עָלֶיךָ רוּחַ ד', וְהִתְנַבִּיתָ עִמָּם; וְנֶהְפַּכְתָּ, לְאִישׁ אַחֵר.

6 And the spirit of the LORD will come mightily upon thee, and thou shalt prophesy with them, and shalt be turned into another man.

This mystical experience according to Idel was practiced by having music induce one into an ecstatic state by Rabbi Abraham Abulafia, who wrote a kabalistic commentary on Maimonides' *Guide for the Perplexed,* in the medieval ages. Music can not only inspire but release the gift of prophecy. However a gemarah in the Talmud qualifies by noting that after the Hurban only a *bat kol* speaks which is likened to the murmering of a dove (Ber 3a) or chirping of a bird (Koheleth Rabbah 7:8) according to Elisha b. Abuya.

Playing on a harp awoke the inspiration that came to Elijah's disciple Elisha (II Kings 3:15) The ability of music to cause a flow of release of the *ruach Hakodesh* is testified in Malakhim Beth, pusek Gimel, verse 15 where Elisha summons a musician to prophecy:

טו וְעַתָּה, קְחוּ-לִי מְנַגֵּן; וְהָיָה כְּנַגֵּן הַמְנַגֵּן, וַתְּהִי עָלָיו יַד-יְהוָה.

15 But now bring me a minstrel.' And it came to pass, when the minstrel played, that the hand of the LORD came upon him.

In Tehillim the harp is called a *Kinnor*, Hebrew: (in Tanakh= kinnor), Greek (in Septuagint= lyre), Latin (in Vulgate= *Psaltrium*), Arabic (in Tafsir= (قيثارة ﻟ Arabic 9 3:4 It was invented by Jubal (Gen. 4:21). Some think the word *kinnor* denotes the whole class of stringed instruments. It was used as an accompaniment to songs of cheerfulness as well as of praise to God (Gen. 31:27; 1 Sam. 16:23; 2 Chr. 20:28; Ps. 33:2; 137:2). In Solomon's time harps were made of almug-trees (1 Kings 10:11, 12). In 1 Chr. 15:21 mention is made of "harps on the Sheminith;" "harps set to the Sheminith;" better perhaps "harps of eight strings." The soothing effect of the music of the harp is referred to in 1 Sam. 16:16, 23; 18:10; 19:9. King David speaks of his harp as an extension of his mind that allows him to clarify and express his thoughts in a way beyond the limits of words:

פִּי, יְדַבֵּר חָכְמוֹת; וְהָגוּת לִבִּי תְבוּנוֹת.	**4** My mouth shall speak wisdom, and the meditation of my heart shall be understanding.
ה אַטֶּה לְמָשָׁל אָזְנִי; אֶפְתַּח בְּכִנּוֹר, חִידָתִי.	**5** I will incline mine ear to a parable; I will open my dark saying upon the harp.

King David is said to have cast out an evil spirit (*ruach rah*) from Saul in melancholy in Shmuel Aleph 16:14-23, where we read:

יֹאמַר-נָא אֲדֹנֵנוּ, עֲבָדֶיךָ לְפָנֶיךָ--יְבַקְשׁוּ, אִישׁ יֹדֵעַ מְנַגֵּן בַּכִּנּוֹר; וְהָיָה, בִּהְיוֹת עָלֶיךָ רוּחַ-אֱלֹהִים רָעָה-- וְנִגֵּן בְּיָדוֹ, וְטוֹב לָךְ. {פ}	**16** Let our lord now command thy servants, that are before thee, to seek out a man who is a skilful player on the harp; and it shall be, when the evil spirit from God cometh upon thee, that he shall play with his hand, and thou shalt be well.'
יז וַיֹּאמֶר שָׁאוּל, אֶל-עֲבָדָיו: רְאוּ-נָא לִי, אִישׁ מֵיטִיב לְנַגֵּן, וַהֲבִיאוֹתֶם, אֵלָי.	**17** And Saul said unto his servants: 'Provide me now a man that can play well, and bring him to me.'
יח וַיַּעַן אֶחָד מֵהַנְּעָרִים וַיֹּאמֶר, הִנֵּה רָאִיתִי בֵּן לְיִשַׁי בֵּית הַלַּחְמִי, יֹדֵעַ נַגֵּן וְגִבּוֹר חַיִל וְאִישׁ מִלְחָמָה וּנְבוֹן דָּבָר, וְאִישׁ תֹּאַר; וַד' עִמּוֹ.	**18** Then answered one of the young men, and said: 'Behold, I have seen a son of Jesse the Beth-lehemite, that is skilful in playing, and a mighty man of valour, and a man of war, and prudent in affairs, and a comely person, and the LORD is with him.'

In Isaiah 30:31 it is said that music can even ensure victory in war:

כִּי-מִקּוֹל ד', יֵחַת אַשּׁוּר; בַּשֵּׁבֶט, יַכֶּה.

31 For through the voice of the LORD shall Asshur be dismayed, the rod with which He smote.

לב וְהָיָה, כֹּל מַעֲבַר מַטֵּה מוּסָדָה, אֲשֶׁר יָנִיחַ ד' עָלָיו, בְּתֻפִּים וּבְכִנֹּרוֹת; וּבְמִלְחֲמוֹת תְּנוּפָה, (נִלְחַם-בה) בָּם.

32 And in every place where the appointed staff shall pass, which the LORD shall lay upon him, it shall be with tabrets and harps; and in battles of wielding will He fight with them.

Ezra and Nehemiah returned to rebuild the Temple and establish its musical temple liturgy in 586 BCE with permission from Cyrus. The importance which music has is shown by the fact that in the original writings of Ezra and Nehemiah a distinction is still drawn between the singers and the Levites (Ezra 2:41, 70; 7: 7, 24; 10: 23; Neh. 7: 44, 73; 10: 29, 40; ad loc.); whereas in the parts of the books of Ezra and Nehemiah belonging to the Chronicles, singers are reckoned among the Levites (Ezra 3: 10; Neh. 11: 22; 12: 8, 24, 27; I Chron. 6: 16). In later 2nd Temple times singers received a priestly position, since Agrippa II bestowed upon them permission to wear the white priestly garment (Josephus, "Ant." 20: 9, 6, Loeb edition). At the dedication of the walls of Jerusalem, Nehemiah formed the Levitical singers into two large choruses, which, after having marched around the city walls in different directions, stood opposite each other at the Temple and sang alternate hymns of praise to God (Neh. 12: 31)

In Chronicles 15 we learn that King David essentially appointed a Levite chorus with accompanying instrumental musicians when we read:

16 And David spoke to the chief of the Levites to appoint their brethren the singers, with instruments of music, psalteries and harps and cymbals, sounding aloud and lifting up the voice with joy.

17 So the Levites appointed Heman the son of Joel; and of his brethren, Asaph the son of Berechiah; and of the sons of Merari their brethren, Ethan the son of Kushaiah;

18 and with them their brethren of the second degree, Zechariah, Ben, and Jaaziel, and Shemiramoth, and Jehiel, and Unni, Eliab, and Benaiah, and Maaseiah, and Mattithiah, and Eliphalehu, and Mikneiah, and Obed-edom, and Jeiel, the doorkeepers.

19 So the singers, Heman, Asaph, and Ethan, [were appointed,] with cymbals of brass to sound aloud;

20 and Zechariah, and Aziel, and Shemiramoth, and Jehiel, and Unni, and Eliab, and Maaseiah, and Benaiah, with psalteries set to Alamoth;

טז וַיֹּאמֶר דָּוִיד, לְשָׂרֵי הַלְוִיִּם, לְהַעֲמִיד אֶת-אֲחֵיהֶם הַמְשֹׁרְרִים, בִּכְלֵי-שִׁיר נְבָלִים וְכִנֹּרוֹת וּמְצִלְתָּיִם--מַשְׁמִיעִים לְהָרִים-בְּקוֹל, לְשִׂמְחָה. {פ}

יז וַיַּעֲמִידוּ הַלְוִיִּם, אֵת הֵימָן בֶּן-יוֹאֵל, וּמִן-אֶחָיו, אָסָף בֶּן-בֶּרֶכְיָהוּ; {ס} וּמִן-בְּנֵי מְרָרִי אֲחֵיהֶם, אֵיתָן בֶּן-קוּשָׁיָהוּ.

יח וְעִמָּהֶם, אֲחֵיהֶם הַמִּשְׁנִים: זְכַרְיָהוּ בֵּן וְיַעֲזִיאֵל וּשְׁמִירָמוֹת וִיחִיאֵל וְעֻנִּי אֱלִיאָב וּבְנָיָהוּ וּמַעֲשֵׂיָהוּ וּמַתִּתְיָהוּ וֶאֱלִיפְלֵהוּ וּמִקְנֵיָהוּ וְעֹבֵד אֱדֹם, וִיעִיאֵל--הַשֹּׁעֲרִים.

יט וְהַמְשֹׁרְרִים, הֵימָן אָסָף וְאֵיתָן--בִּמְצִלְתַּיִם נְחֹשֶׁת, לְהַשְׁמִיעַ.

כ וּזְכַרְיָה וַעֲזִיאֵל וּשְׁמִירָמוֹת וִיחִיאֵל, וְעֻנִּי וֶאֱלִיאָב, וּמַעֲשֵׂיָהוּ, וּבְנָיָהוּ--בִּנְבָלִים, עַל-עֲלָמוֹת.

כא וּמַתִּתְיָהוּ וֶאֱלִיפָלֵהוּ, וּמִקְנֵיָהוּ וְעֹבֵד אֱדֹם, וִיעִיאֵל, וַעֲזַזְיָהוּ--בְּכִנֹּרוֹת עַל-הַשְּׁמִינִית, לְנַצֵּחַ.

21 and Mattithiah, and Eliphalehu, and Mikneiahu, and Obed-edom, and Jeiel, and Azaziah, with harps on the Sheminith, to lead.

כב וּכְנַנְיָהוּ שַׂר-הַלְוִיִּם, בְּמַשָּׂא--יָסֹר, בַּמַּשָּׂא, כִּי מֵבִין, הוּא.

22 And Chenaniah, chief of the Levites, was over the song; he was master in the song, because he was skilful.

In I Chronicles 9:33 we had been given genealogies of these musician families ad loc the verses:

וְאֵלֶּה הַמְשֹׁרְרִים רָאשֵׁי אָבוֹת לַלְוִיִּם, בַּלְּשָׁכֹת--פטירים (פְּטוּרִים): כִּי-יוֹמָם וָלַיְלָה עֲלֵיהֶם, בַּמְּלָאכָה.

33 And these are the singers, heads of fathers' houses of the Levites, who dwelt in the chambers and were free from other service; for they were employed in their work day and night.

In I Chronicles 16 4-6 elaboration of the orchestra is specified further:

וַיִּתֵּן לִפְנֵי אֲרוֹן יְהוָה, מִן-הַלְוִיִּם--מְשָׁרְתִים; וּלְהַזְכִּיר וּלְהוֹדוֹת וּלְהַלֵּל, לַיהוָה אֱלֹהֵי יִשְׂרָאֵל. {ס}

4 And he appointed certain of the Levites to minister before the ark of the LORD, and to celebrate and to thank and praise the LORD, the God of Israel:

5 Asaph the chief, and second to him Zechariah, Jeiel, and Shemiramoth, and Jehiel, and Mattithiah, and Eliab, and Benaiah, and Obed-edom, and Jeiel, with psalteries and with harps; and Asaph with cymbals, sounding aloud;

ה אָסָף הָרֹאשׁ, וּמִשְׁנֵהוּ זְכַרְיָה; יְעִיאֵל וּשְׁמִירָמוֹת וִיחִיאֵל וּמַתִּתְיָה וֶאֱלִיאָב וּבְנָיָהוּ וְעֹבֵד אֱדֹם וִיעִיאֵל, בִּכְלֵי נְבָלִים וּבְכִנֹּרוֹת, וְאָסָף, בַּמְצִלְתַּיִם מַשְׁמִיעַ.

6 and Benaiah and Jahaziel the priests with trumpets continually, before the ark of the covenant of God.

ו וּבְנָיָהוּ וְיַחֲזִיאֵל, הַכֹּהֲנִים--בַּחֲצֹצְרוֹת תָּמִיד, לִפְנֵי אֲרוֹן בְּרִית-הָאֱלֹהִים.

In Chronicles 16 verses 39-41 we reader further:

39 and Zadok the priest, and his brethren the priests, before the tabernacle of the LORD in the high place that was at Gibeon,

וְאֵת צָדוֹק הַכֹּהֵן, וְאֶחָיו הַכֹּהֲנִים, לִפְנֵי, מִשְׁכַּן ד-- בַּבָּמָה, אֲשֶׁר בְּגִבְעוֹן.

40 to offer burnt-offerings unto the LORD upon the altar of burnt-offering continually morning and evening, even according to all that is written in the Law of the LORD, which He commanded unto Israel;

מ לְהַעֲלוֹת עֹלוֹת לד' עַל-מִזְבַּח ח הָעֹלָה, תָּמִיד-- לַ בֹּ קֶ ר וְ לָ עָ רֶ ב; וּלְכָל-הַכָּתוּב בְּתוֹרַת ד', אֲשֶׁר צִוָּה עַל-יִשְׂרָאֵל.

41 and with them Heman and Jeduthun, and the rest that were chosen, who were mentioned by name, to give thanks to the LORD, because His mercy endureth for ever;

מא וְעִמָּהֶם, הֵימָן וִידוּתוּן, וּשְׁאָר הַבְּרוּרִים, אֲשֶׁר נִקְּבוּ בְּשֵׁמוֹת--לד', כִּי לְעוֹלָם, חַסְדּוֹ.

מב וְעִמָּהֶם הֵימָן וִידוּתוּן **42** and with them Heman and
חֲצֹצְרוֹת וּמְצִלְתַּיִם , Jeduthun, to sound aloud with
לְמַשְׁמִיעִים, וּכְלֵי, שִׁיר trumpets and cymbals, and with
הָאֱלֹהִים; וּבְנֵי יְדוּתוּן, instruments for the songs of God;
לַשָּׁעַר. and the sons of Jeduthun to be at the
gate.

Heman and Juduthun play percussion instruments of
trumpets and cymbals. The horned instruments previously
were played by Benaiah and Jahaziel the priests with
trumpets. By listing first the stringed instruments there
seems to be a hierarchy to the estimation of the nobility of
the instruments. We had learned in verse 5 that the
stringed instruments were played by Asaph the chief, and
second to him Zechariah, Jeiel, and Shemiramoth, and
Jehiel, and Mattithiah, and Eliab, and Benaiah, and Obed-
edom, and Jeiel, with psalteries and with harps; Yet
Chronicles is not finished yet with musical history. We
again see the theme of music linked with prophecy in
Chronicles 25 1-7:

וַיַּבְדֵּל דָּוִיד וְשָׂרֵי הַצָּבָא לַעֲבֹדָה, **1** Moreover David and the
לִבְנֵי אָסָף וְהֵימָן וִידוּתוּן, הַנְּבִיאִים captains of the host
(הַנִּבְּאִים) בְּכִנֹּרוֹת בִּנְבָלִים, separated for the service
וּבִמְצִלְתָּיִם; וַיְהִי, מִסְפָּרָם, אַנְשֵׁי certain of the sons of
מְלָאכָה, לַעֲבֹדָתָם. Asaph, and of Heman, and
of Jeduthun, who should
prophesy with harps, with
psalteries, and with
cymbals; and the number of
them that did the work
according to their service
was:

2 of the sons of Asaph: Zaccur, and Joseph, and Nethaniah, and Asarelah, the sons of Asaph; under the hand of Asaph, who prophesied according to the direction of the king.

3 Of Jeduthun: the sons of Jeduthun: Gedaliah, and Zeri, and Jeshaiah, Hashabiah, and Mattithiah, six; under the hands of their father Jeduthun with the harp, who prophesied in giving thanks and praising the LORD.

4 Of Heman: the sons of H e m a n : Bukkiah, Mattaniah, Uzziel, Shebuel, and Jerimoth, Hananiah, Hanani, Eliathah, Giddalti, and Romamti-ezer, Joshbekashah, Mallothi, Hothir, Mahazioth;

5 all these were the sons of Heman the king's seer in the things pertaining to God, to lift up the horn. And God gave to Heman fourteen sons and three daughters.

ב לִבְנֵי אָסָף, זַכּוּר וְיוֹסֵף וּנְתַנְיָה וַאֲשַׂרְאֵלָה--בְּנֵי אָסָף: עַל, יַד-אָסָף, הַנִּבָּא, עַל-יְדֵי הַמֶּלֶךְ.

ג לִידוּתוּן--בְּנֵי יְדוּתוּן גְּדַלְיָהוּ וּצְרִי וִישַׁעְיָהוּ חֲשַׁבְיָהוּ וּמַתִּתְיָהוּ שִׁשָּׁה, עַל יְדֵי אֲבִיהֶם יְדוּתוּן בַּכִּנּוֹר, הַנִּבָּא, עַל-הֹדוֹת וְהַלֵּל לד'. {ס}

ד לְהֵימָן--בְּנֵי הֵימָן בֻּקִּיָּהוּ מַתַּנְיָהוּ עֻזִּיאֵל שְׁבוּאֵל וִירִימוֹת חֲנַנְיָה חֲנָנִי, אֱלִיאָתָה גִדַּלְתִּי וְרֹמַמְתִּי עֶזֶר, יָשְׁבְּקָשָׁה מַלּוֹתִי, הוֹתִיר מַחֲזִיאוֹת.

ה כָּל-אֵלֶּה בָנִים לְהֵימָן, חֹזֵה הַמֶּלֶךְ בְּדִבְרֵי הָאֱלֹהִים--לְהָרִים קָרֶן; וַיִּתֵּן הָאֱלֹהִים לְהֵימָן, בָּנִים אַרְבָּעָה עָשָׂר-- וּבָנוֹת שָׁלוֹשׁ.

6 All these were under the hands of their fathers for song in the house of the LORD, with cymbals, psalteries, and harps, for the service of the house of God, according to the direction of the king-- Asaph, Jeduthun, and Heman.

ו כָּל-אֵלֶּה עַל-יְדֵי אֲבִיהֶם בַּשִּׁיר בֵּית ד', בִּמְצִלְתַּיִם נְבָלִים וְכִנֹּרוֹת, לַעֲבֹדַת בֵּית הָאֱלֹהִים--עַל יְדֵי הַמֶּלֶךְ, אָסָף וִידוּתוּן וְהֵימָן.

7 And the number of them, with their brethren that were instructed in singing unto the LORD, even all that were skilful, was two hundred fourscore and eight.

ז וַיְהִי מִסְפָּרָם עִם-אֲחֵיהֶם, מְלֻמְּדֵי-שִׁיר לד': כָּל-הַמֵּבִין--מָאתַיִם, שְׁמוֹנִים וּשְׁמוֹנָה.

I Chronicles is not the only source for our knowledge of the musicality in the Beit HaMikdash. In II Chronciles 29 we read further:

25 And he set the Levites in the house of the LORD with cymbals, with psalteries, and with harps, according to the commandment of David, and of Gad the king's seer, and Nathan the prophet; for the commandment was of the LORD by His prophets.

כה וַיַּעֲמֵד אֶת-הַלְוִיִּם בֵּית ד', בִּמְצִלְתַּיִם בִּנְבָלִים וּבְכִנֹּרוֹת, בְּמִצְוַת דָּוִיד וְגָד חֹזֵה-הַמֶּלֶךְ, וְנָתָן הַנָּבִיא: כִּי בְיַד-ד' הַמִּצְוָה, בְּיַד-נְבִיאָיו.

כו   וַיַּעַמְדוּ הַלְוִיִּם בִּכְלֵי   26 And the Levites stood with the
דָוִיד, וְהַכֹּהֲנִים בַּחֲצֹצְרוֹת.   instruments of David, and the
{פ}   priests with the trumpets.

כז   וַיֹּאמֶר, חִזְקִיָּהוּ,   27 And Hezekiah commanded to
לְהַעֲלוֹת הָעֹלָה, לְהַמִּזְבֵּחַ;   offer the burnt-offering upon the
וּבְעֵת הֵחֵל הָעוֹלָה, הֵחֵל   altar. And when the burnt-offering
שִׁיר-ד׳ וְהַחֲצֹצְרוֹת,   began, the song of the LORD began
וְעַל-יְדֵי, כְּלֵי דָוִיד   also, and the trumpets, together
מֶלֶךְ-יִשְׂרָאֵל.   with the instruments of David king
of Israel.

As noted earlier, King David is said to have cured the
melancholy of Saul by playing the harp. Dr. Joachin Braun
has noted the differences in size of musical period
instruments in antiquity such as the harp during the 1st
temple (*kinur*), 2nd temple (*lyre*), and Roman and
Byzantine periods (*psaltrium*) whereby the size, pitch, and
harmony of the instrument was changed by differences in
the representations of the instrument in these different
time periods.

Likewise for example the "*taf*" or tambourine employed by
Miriam who took the women out in song and dance in
parasha *shirat ha-yam* after the parting of the Reed Sea
(Shemot 14) later in history in the 16th century in Ladino
translations of the Tanakh is translated as "Castenet." We
thus can view the following description of Miriam after
Kriat Yam Suf as a musical refrain:

וַתִּקַּח מִרְיָם הַנְּבִיאָה אֲחוֹת   20 And Miriam the prophetess, the
אַהֲרֹן, אֶת-הַתֹּף--בְּיָדָהּ;   sister of Aaron, took a timbrel in
וַתֵּצֶאןָ כָל-הַנָּשִׁים אַחֲרֶיהָ,   her hand; and all the women went
בְּתֻפִּים וּבִמְחֹלֹת.   out after her with timbrels and
with dances.

וַתַּעַן לָהֶם, מִרְיָם: **כא** 21 And Miriam sang unto them:
שִׁירוּ לד' כִּי-גָאֹה גָּאָה, סוּס Sing ye to the LORD, for He is
וְרֹכְבוֹ רָמָה בַיָּם. highly exalted: the horse and his
rider hath He thrown into the sea

Miriam's song echoes Moshe's intonation to song previously when the chief of the prophets proclaims:

אָז יָשִׁיר-מֹשֶׁה וּבְנֵי יִשְׂרָאֵל **א** 1 Then sang Moses and the
אֶת-הַשִּׁירָה הַזֹּאת, לד', וַיֹּאמְרוּ, children of Israel this song
לֵאמֹר: אָשִׁירָה לד' כִּי-גָאֹה גָּאָה, unto the LORD, and spoke,
סוּס וְרֹכְבוֹ רָמָה בַיָּם. saying: I will sing unto the
LORD, for He is highly
exalted; the horse and his
rider hath He thrown into
the sea.

עָזִּי וְזִמְרָת יָהּ, וַיְהִי-לִי לִישׁוּעָה; **ב** 2 The LORD is my strength
זֶה אֵלִי וְאַנְוֵהוּ, אֱלֹהֵי אָבִי וַאֲרֹמְמֶנְהוּ. and song, and He is
become my salvation; this
is my God, and I will
glorify Him; my father's
God, and I will exalt Him.

This echo effect is indeed in parasha *Beshallach* accompanied by song of a special trope as is the haftorah of Deborah.

In the course of my graduate studies various rabbis who ran the minyanim I davoned in, would ask me to do little research projects. One rabbi sought material on the example of the *magrepha,* a strictly 2nd Temple instrument which is mentioned in the mishnah said to be shaped like a shovel and some sort of percussion instrument.

In the Mishnah of Tamid ה משנה מסכת תמיד פרק we learn of an instrument, a rake-like or shovel vessel used for brushing away the ashes from the altar, or a type of musical percussion instrument), and thrown down to the floor!

הגיעו בין האולם ולמזבח נטל אחד את המגרפה וזורקה בין האולם ולמזבח אין אדם שומע קול חברו בירושלים מקול המגרפה ושלשה דברים היתה משמשת כהן ששומע את קולה יודע שאחיו הכהנים נכנסים להשתחוות והוא רץ ובא ובן לוי שהוא שומע את קולה יודע שאחיו הלוים נכנסים לדבר בשיר והוא רץ ובא וראש המעמד היה מעמיד את הטמאים בשער המזרח:

In the language of the Mishna (Tamid 5, 6) "the sound made by the *magrepha* falling was so deafening at that moment that in the entire city of Jerusalem, no one could hear his friend speaking!"

As noted before the *magrepha* was a 2nd temple Instrument.

The instruments of the 1st Temple (*Bayit Rishon*) are given to us as being instruments (*kelai zemer*) gleaned from the Biblical text, particularly Psalms (Tehillim):

i. Kinur כנור

ii. Ugav עוגב

iii. Taf תף

iv. Shofar שופר

v. tsël-tsë-lim צלצלים

vi. chä-lel חליל

vii. ma-tsel-tä'-yem מצילתים

viii. ma-nä-än-em מננים

ix. ba-khol' ä-tse' va-ro-shem

x. chä-tsots-rot חצוצרות

xi. ba-khol' ä-tse' va-ro-shem

xii. magrepha (2nd temple instrument only) מגרפה

xiii. etc.

A website of the Temple Institute[1] sees this incident of the *magrepha* as serving 3 functions:

> 1). When the priests who were outside the court heard the sound, they knew that their colleagues within were about to prostrate themselves before the Divine Presence... and they ran to bow down with them.
>
> 2). When the Levites heard it, they knew that the Levite choir was about to enter the court and stand upon the platform, to begin their service of the daily song. They too, ran to join their brothers...
>
> 3). And when the Assembly Head (the official in charge of the Israelites who stand in the Temple to accompany the sacrificial service, as representatives of the entire nation) heard, he separated the priests who had become defiled, and stood them all together in the Eastern Gate. This way, everyone could see that they were impure and therefore could not serve in the Holy Temple, and no one would suspect that they had any other reason for not participating in the service.

Whatever the reason for throwing down the *magrepha* it is significant that its sound was so powerful and intense. It may remind one of the power of the *keturet* offered in the *Beit HaMikdash* that according to another Mishnah was said to waft as far as the galilee so that the goats there

---

[1]   www.templeinstitute.org/day_in_life/magrepha.htm,   accessed February 11, 2018.

would sneeze! In the Roman period of the second temple, of which the magrepha is a unique example the article by Emmanuel Friedheimn, (titled *Jewish society in the Land of Israel and the challenge of music in the Roman period*, which appeared in 2012 in *Review of Rabbinic Judaism - Ancient, Medieval, and Modern* 15,1 (2012) 61-88, sets out a call for much more work to be done in this epoch of Jewish music so close to the *Hurban*.

From my father's love of music I came to feel and experience music's magical ability to transport, and how it represents the spirit of a people. According to the rabbis of the Talmud "one had never seen a beautiful sacred building if they had not gazed upon the *Beit HaMikdash*, and one never heard "sublime music" if they never heard the Levites singing on the steps, the sweet singers of the Temple liturgies. Indeed Josephus tells us that the *Beit Hamikdash* was made of marble that was painted Mediterranean sea blue, and when the sun rebounded off of the gold dome, the blue painted marble had the appearance of shimmering Mediterranean pure sea waters. Thus Rabbi Akiva in *Maseket Hagigah* says, "*al tamru mayim mayim*" because the shimmering appearance of the illusory play of light on the blue marble is not indeed water, but marble painted blue in the sunlight. So too the music of the *Beit HaMikdash* was even more holy and we learn many insights about this music from the Talmudim. One remark notes that when the Levites were taken in slavery to Babylon one Levitical family that excelled in instrumental music cut off their thumbs so that they not be able to teach music to their captors (*Pesikta Rabbasi* 31). "On the rivers of Babylon there we sat and wept, when we remembered Zion" for the Levites refused to sing a new song of an old land in a land of captivity in deference to the music of the *Beit HaMikdash* (Ps 137).

Rashi comments that tolal is a musical instrument, with regards to the verses:

ב עַל-עֲרָבִים בְּתוֹכָהּ--   2 Upon the willows in the midst
תָּלִינוּ, כִּנֹּרוֹתֵינוּ.   thereof we hanged up our harps.

ג כִּי שָׁם שְׁאֵלוּנוּ שׁוֹבֵינוּ,   3 For there they that led us captive
דִּבְרֵי-שִׁיר-- וְתוֹלָלֵינוּ   asked of us words of song, and our
שִׂמְחָה: tormentors asked of us mirth:
שִׁירוּ לָנוּ, מִשִּׁיר צִיּוֹן.   'Sing us one of the songs of Zion.'

Radak holds that this music instrument *tolal* is a harp, *kinor*, rooted in the word hanging because the Levites hung the instrument on willows, a tree representing grief and mourning. Radak also relates *tolal* to weeping, our cruel captors demanded that instead of weeping we should display joy. *Pesikta Rabbasi* 28 notes that when the Levites refused to sing songs before the idols of Babylon, their cruel captors were angry, and slew multitudes and piled up (*talim*) in mounds. Thus the reading "despite our mounds of murdered victim bodies" we rejoiced over our decision to resist singing to their idols. The Malbim explains the cruelty of the Babylonians demand that while the Jews used to sing wonderful songs about Zion and the Temple music, now it is demanded that the Jews forget Zion and accept Babylon as the homeland and sing to Babylon, the same praises once sang to Zion claiming it the most wonderful place "the consummation of all beauty" (Ps 50:2). When the Levite instrumentalists were demanded to serenade Nebuchadnezzar the Levite instrumentalists immediately without hesitation cut off their thumbs maiming their ability to play stringed instruments. They thus did not refuse to play, they subversively were able to reply, "How can we sing the song of Hashem? We cannot make any more sublime music with these crippled hands (*Pesikta Rabbasi* 31).

Rabbinic Thought

Indeed in the rishonim period Rabbinic *halakhah* proscribed the use of music instruments even at weddings in order to remember the true and final place for sacred music to be sung by the Levites in the Temple as expressions of kedushah and spiritual elevation. Joshua Jacobson article, *"We hung up our harps : rabbinic restrictions on Jewish music "* that appeared in the *Journal of Synagogue Music* 25,2 (1998) 33-53 in 1998 makes this clear. Cyrus Adler (see JE) however suggests that post-Hurban these restrictions were often not followed or generally heeded when it became a question of song in worship (comp. Giṭ. 7a; Soṭah 48a; Alfasi on Ber. 25b; Asheri on Ber. 30b; Shulḥan 'Aruk, Oraḥ Ḥayyim, 560, 3). Adler supports his position by noting That in synagogues the *sheliach zibur* was required to have a pleasant voice and a clear enunciation (Ta'an. 16a; Pesiḳ. R. 25 [ed. Friedmann, p. 127a]; comp. Meg. 24b, 32a; Yer. Sheḳ. 1; Yalḳ., Prov. 932). Further the voluntary assistance of good vocalists was regarded as meritorious. Among such Ḥiyya bar Adda is prominently mentioned (comp. Pesiḳ. 97a). Adler continues, "The Shema', known to all, was chanted in unison; but the *"Tefillah"* (Shemoneh 'Esreh) was intoned by the officiant only, the congregation responding loudly in unison, as also when *Ḳaddish* was read (Soṭah 49a; Shab. 119b). The Psalms were chanted originally in a responsive antiphony (Soṭah 30b; comp. Graetz in "Monatsschrift," 1879, p. 197); but soon the antiphony developed into a general unison, as became the case, too, with the other passages gradually added to the ritual (Cant. R. 27a, end; Rashi on Ber. 6a; but comp. Zunz, "S. P." p. 61); JE)

Yet a number of rabbis embraced the wonderous power and importance of music in conveying rabbinic theology. Eric Werner set the ground for this study, *Theologie der Musik im frühen Judentum*, which appeared in 1985 in Zeitschrift für Religions- und Geistesgeschichte 37,3 (1985) 258-260. Enrico Fubini in Italy has also explored this in his essay *La musica nella tradizione biblica and talmudica*, which appeared in 1991. Also see for Hebrew texts on Andalus, I. Adler *Hebrew Writings Concerning Music in Manuscripts and Printed Books From Geonic Times up to 1800* (München, 1975).

Rav Saadia Gaon is one rabbi who wrote on music. Shiloah Amnon in his article *Musical concepts in the works of Saadia Gaon*, which was published in 2004 in Aleph 4 (2004) 265-282 notes Rav Saadia's great appreciation of a philosophy of music. Ufl Hazen also in his essay *Saadya Gaon on music - melody or rhythm?* Which appeared in 1999 in Jewish Studies at the Turn of the Twentieth Century I (1999) 406-413 continues our knowledge of Rav Saadia's musical appreciation. Hanoch Avenary brings to our attention from the important find of the Cairo Geniza in his article *A Geniza find of Saadya's Psalm-preface and its musical aspects* which appeared in HUCA in 1968 further knowledge. See also Adler, I., *Hebrew Writings Concerning Music in Manuscripts and Printed Books from Geonic Times up to 1800* (Muenchen, 1975).

Rav Saadia analyzes arrangements of musical notes although the western mode of musical notation with a treble cleff is from the 16th century of Bach and Buxtehude. The medievals had however their own sophisticated modes of musical notation at least dating back to the Masorites in Tiberia who fixed and stabilized the cantillation of the trope.

The Hebrew term *Niggun* means "tune" and when its melody is primarily in view, by the Judæo-German term "*steiger*" (scale). When its modal peculiarities and tonality are under consideration, then by the Romance word "gust" and the Slavonic "*skarbowa*" when the taste or style of the rendering especially marks it off from other music. The use of these terms, in addition to such less definite Hebraisms as "*ne'imah*" ("melody"), show that the scales and intervals of such prayer-motives have long been recognized (*e.g.*, by Saadia Gaon in the tenth century; comp. end of *Sefer Emunot we-Deot*). Ancient tradition, from the days when the Jews who passed the Middle Ages in Teutonic lands were still under the same tonal influences as the peoples in southeastern Europe and Asia Minor yet are, chromatic scales (*i.e.*, those showing some successive intervals greater than two semitones).

Musical Syncretism

Yet the Jews in the diaspora repeatedly adapted and borrowed musical melodies from their host cultures but made them distinctively Jewish. The "*harmonia*," or manner in which the prayer-motive will be amplified into *hazzanut*, is sometimes measured by syncretism of surrounding musical traditions. Thus in medieval Ashkenaz ,Troubadours, *trouvères*, and minnesingers, as well as jongleurs and minstrels, had by this time laid the foundations of modern melody in their ever-extending use of the diatonic scale.

In medieval Sephardic culture Simon Duran (1361–1444) notes that the synagogue *Hazanim* took into the synagogue some gentile tunes so that the number of Neo-Hebraic hymns rapidly increased because of their pleasing jingle often, their tender expressiveness sometimes, early (comp. Simon Duran, "*Magen Abot*," 52b) which led to their retention and perpetuation and to their adoption as the traditional setting of the verses.

Abraham ibn Ezra (on Ps. 8) refers to the introduction of alien airs in the eleventh century; and according to S. Archevolti in the sixteenth century ("'*Arugat ha-Bosem*," p. 100)- the practice was a general one in the days of Judah ha-Levi (early part of 12th cent.).

Musical Syncretism Continued

Later in Italy thus Shlomo de Rossi adopted musical innovations of the Renaissance baroque music that he heard and applied them to his Hebrew choral music. Likewise Sephardic music often adopts certain musical aspects of traditions in Arabic music. Under the influence of Slavonic and Gipsy passion in melody, or in Moslem lands, where the short, infinitely repeated phrase in the distinctive Perso-Arab scales still prevails in every-day music, Jews picked up techniques from other musical traditions. According to Francis Cohen (JE) in the article *Lekah Dodi* (see Jewish Encyclopedia, 7: 676) three such melodies adopted from popular use into Jewish worship— one, Moorish, of the tenth; one, Polish, of the sixteenth; and one, German, of the seventeenth century. The well-known melody of *Ma'oz Ẓur* was likewise adapted from a street.

Japhet's collection (Frankfort-on-the-Main, 1855; 3d ed. 1903, No. 60) of the synagogal melodies of southwest Germany, which are particularly replete with folk-song elements. Much of Eastern European Ashkenaz music from the 18th century on, incorporates a considerable mass of melody directly adapted from the folk-song of Gentile neighbors. Cantors have always kept an ear to the ground for great music. This is evident in the use of certain modal inflections, cadence patterns, and styles of vocal ornamentation that can be traced to Russian, Greek, and Armenian chants, as well as to secular folk songs.

The YIVO Encyclopedia and Jewish Encyclopedia have a good synopsis of the developments of *Hazzanut* in Eastern Europe along with other articles of music in Ashkenazic Jews such as 1) music for sacred etc., 2) Hasidism, music 3) music, study of, 4) Jewish folk music, 5) music.[2]

In 1660 Ḥayim Zelig of Lwów became the cantor in Fürth. Several East European cantors were employed by the Jewish communities in Amsterdam, including Yeḥi'el Mikha'el (ca. 1700) and Leib Elyakim (1730). In the wealthy city of Pressburg (now Bratislava), the German influence was more strongly felt, while the northeastern portion of Hungary absorbed Polish musical traditions. Dovidl Brod Strelisker (1783–1848) had no formal musical training and did not study with a cantor. Yet he had talent and voluntarily started singing in synagogues during his travels. Ultimately he served as a cantor in Pest, exerting enormous influence over cantorial practices in Hungary. Nissl Belzer (1824–1906), who was one of the most gifted composers and choir leaders of the nineteenth century. Zeydl Rovner (born Yankev Shmuel Maragowski) and Nissl Belzer (Nisan Spivak), who were known both for their solo recitatives and choral compositions. Spivak in Polish means Cantor.

East European cantors who published important collections include A. Dunajewski (1843–1911), Eli'ezer Gerovich (1844–1914) and Barukh Schorr (1823–1904). From 1881 to 1891, the *Österreichisch-Ungarische Cantoren-Zeitung* (Austro-Hungarian Cantors' Magazine) provided a guide to Jewish music and professional life in Austria and

---

[2] YIVO encyclopedia: http://www.yivoencyclopedia.org/ and Jewish Encyclopedia (1906): http://www.jewishencyclopedia.com/

Hungary. In Częstochowa, Avrom Ber Birnboym printed several volumes of liturgical music for Shabbos and the Yamim Noraim, but also published a journal for cantors, *Yarḥon ha-ḥazanim* (Cantors' Monthly; 1896–1897)

Cities such as Vilna, Czestochowa, Warsaw, Budpest, Odessa, Pressubrg, Lwow, Kishinev, Berdichev could afford important cantors Cantor Shmuel Vigoda (1894–1990) wrote an important anecdotal history of Hazanut, titled *Legendary Voices: The Fascinating Lives of the Great Cantors;* [1981), in which he contrasts the more insular practices in the smaller community of Berdichev with the more cosmopolitan musical life of synagogues in Odessa.

In the east, the Hasidic movements influenced musical styles both inside and outside the synagogue, placing a greater emphasis on the relationship between wordless melody (nigunim) and the mystical, ecstatic experience that was manifest as well in the regular pattern of repetition that structured many Hasidic *nigunim.*

Even amongst the Haredi an example of a conscious adaptation of a non-Jewish melody is the incorporation of the Napoleonic march for its remarkable joyous, rhythmic character. It was played in 1812 by the armies of Napoleon when they crossed the border near Prussia in their invasion of Russia. The *Alter Rebbe* had left his native town of Liadi when the armies of the enemy were approaching. He asked that the march be sung for him and, after a moment's consideration, designated the march as a song of victory. This is ironic because the *Baal haTanya* opposed Napoleon's modern Enlightenment policy to assimilate the Jews as secularized French citizens. It is traditional with Lubavitcher *Chassidim* to sing Napoleon's march at the conclusion of the *Ne'ilah* service on *Yom Kippur,* before the sounding of the *Shofar.* They thus subversively adopt the melody for Hassidus. The singing of this *melody* symbolizes the victory of the Jewish people over *"the accuser"* and that their prayers have been accepted and they are assured of a Happy New Year.

Music Expresses Whole Gamut of Emotions from Grief to Joy

On *Tisha b'av* the book of Lamentations (Eikah) is chanted in a minor key representative of dirges in musicology to express the deep emotion of and feeling of ultimate loss caused by the *Hurban*. David laments Yonathan and Saul in Samuel I, 18 with instructing the sons of Judah to learn the bow:

וַיֹּאמֶר לְלַמֵּד בְּנֵי יְהוּדָה קָשֶׁת הִנֵּה כְתוּבָה עַל סֵפֶר הַיָּשָׁר

Yonathan and Saul who were slain on Har Gilboa are referred to as the beauty of Israel and "how the mighty have fallen," some of the most sublime poetry ever written:

הַצְּבִי יִשְׂרָאֵל עַל בָּמוֹתֶיךָ חָלָל אֵיךְ נָפְלוּ גִבּוֹרִים

Thus music can express the apex of joy and yet the depths of sorrow. When King David brings the Ark of Hashem from Hebron to Jerusalem, reigning 7 years in Hebron and 33 in Jerusalem, David accompanies the procession in song and dance expressing the greatest joy to Hashem:

בְּכֹל עֲצֵי בְרוֹשִׁים וּבְכִנֹּרוֹת ד׳ וְדָוִד וְכָל בֵּית יִשְׂרָאֵל מְשַׂחֲקִים לִפְנֵי וּבִנְבָלִים וּבְתֻפִּים וּבִמְנַעַנְעִים וּבְצֶלְצֶלִים

And David and all the house of Israel played before the LORD with all manner of instruments made of cypress-wood, and with harps, and with psalteries, and with timbrels, and with sistra, and with cymbals. Nothing could be more joyous. When Michal rebukes David for dancing before the riffraff, lowering his royal dignity like a commoner, David corrects her by telling her he danced in joy before Hashem:

אֲשֶׁר בָּחַר בִּי מֵאָבִיךְ וּמִכָּל בֵּיתוֹ ד' וַיֹּאמֶר דָּוִד אֶל מִיכַל לִפְנֵי
ד' עַל יִשְׂרָאֵל וְשִׂחַקְתִּי לִפְנֵי ד' לְצַות אֹתִי נָגִיד עַל.

Psalms during the 1st and 2nd Temple periods were set to music. We know this by many musical instructions innate to the language of Psalms such as "selah" which indicates an ascending scale on a stringed instrument. Emil Hirsch points out that in II Chron. 5: 13 that at the dedication of the Temple the playing of the instruments, the singing of the Psalms, and the blare of the trumpets sounded as one sound, suggesting the knowledge of octaves. Probably the unison of the singing of Psalms was the accord of two voices an octave apart. This explains the terms "'al 'alamot" and "'al ha-sheminit." As late as the Geonic period the Hai Gaon notes that (d. 1038; comp. Zunz, "Ritus," p. 11) (see Jew. Encyc. iii. 539, s.v. Cantillation) the intonations accompanying Tehillim recital to be very ancient, and possibly to date back to the method of rendition utilized for the Psalms in the Temple.

Sol Finesinger has written an excellent article titled, "Musical Instruments in the Tanakh" in which the scholar notes the differences in language for musical instruments mentioned in the Tanakh as they appear in their latter translations in the Targumim (Aramaic), Septuagint (Greek), Vulgate (Latin), Tafsir (Arabic), Ladino (Judeo-Spanish), Le texte sacre (French), Die Heilige Schriften und Ihren Verdeutschung (German) etc. Finesinger's scholarship is emblematic of the Wissenschaft des Judentums beweigung (Science of Judaism Movement) of the 19th century in German. Many scholarly works on Biblcial music appeared in 19th century Germany including Saalschütz, Gesch. und Würdigung der Musik bei den Alten Hebräern, (1829) and Forkel's All-gemeine Gesch. der Musik.

Rabbi Joshua ben Hananiah, who had served in the sanctuary as a member of the Levitical choir ('Ar. 11b), told how the Temple choristers went in a body to the synagogue from the orchestra by the altar (Suk. 53a), and so participated in both services. As the part of the instruments in the Temple musical ensemble was purely that of accompaniment, and the voices gave a strong rendition without accompaniment (comp. Suk. 50b *et seq.*; 'Ar. 11; Num. R. vi.). In fact a sugya attests that the success of one Levitical singer in reciting the Hallel in the Temple was his strong voice that seemed to appear to make the roof beams tremble. Rambam and Ramban in each of their *Sefer Ha-Mitzvot* differ on whether the recitation of Hallel is doereita or derabbanan? The Tashbaz in *sefer Zohar ha-Rakiah* tries to reconcile this point of difference (*makhloket*).

Hasidism and Music

All periods of Jewish history gave rise to expressions of different cultures of Jewish music from Ladino songs, Sephardic pisgomim, to Yiddish folk songs. These music traditions are rich in beauty and deep in expressing the whole gamut of the emotions from sadness to ultimate joy and happiness. "Music is the food of love" indeed, because it is the opposite of anything corporeal, but wholly an expression of the eternal spirit of human beings striving to reach the stars (*ad astra*), and G-dliness. Maimonides wrote a *teshuvah* on "listening to music" and from this teshuvah we learn that the Rambam distinguished between good and bad music. Bad music is rude, vulgar, and a thing of common people who live at the level of animals deriving no pleasure except from eating and things of the body while on the other hand good music serves a number of functions including, lifting one's spirits from melancholy i.e. a therapeutic function, and ultimately giving expression to the ultimate longing and desire of the human being "for dwelling closer to Hashem.

Benjamin Mintz' 1930 Hebrew *Sefer haHistalkut* gives accounts of 42 Hasidic *Tsadikim* between 1760 and 1904 depicting their departures in passing from *olam ha-zeh* to *olam ha-bah*. The whole Hebrew text can be found at www.daat.ac.il/daat/history/hasidut/sefer-2.htm. Mintz selection of 42 death bed departures of *tzadikim* is based on *parasha Masei* that recounts 42 departure points or way stations of the people Israel in the dessert until reaching the promised land in Israel. As an allegory this suggests that striving for the Edenic promised land is only finally possibly obtained depending on the merits of the soul in the afterlife. In at least two accounts of this Hasidic legends work on last moments of Hasidic Rebbes, music plays a key role.

A Hasidic rebbe is said to have been so influenced by music's power that the Rebbe at *seudah shelishit,* when various Hebrew songs are sung, actually practiced a mystical technique of allowing his soul in musical ecstasy and rapture to "go out and not return" (*yotzei ve bili teshuv*) and the Rebbe song a *niggun* on particular shabbos *seudah shelishit* with all devotion to Hashem in *ahavas Hashem, ahavas Torah,* and *ahavas olam,* that the Rebbe's soul did indeed go out and depart to the higher realms. Another Hasidic legendary story collected by Joseph Mintz in *Sefer Hishtalkut* (the book of soul Journeys) recalls a different Hasidic rebbe, The elder of Ujhel, Rabbi Moshe Teitelbaum. Rabbi Teitelbaum's last words were to be granted to hear David's mystical harp make music from the breeze (maseket Berachot 3b). The source is found in Tehillim 119:62:

עַל מִשְׁפְּטֵי צִדְקֶךָ: חֲצוֹת־לַיְלָה אָקוּם לְהוֹדוֹת לָךְ

This origin of the ritual of *tikun hazot* echoes in the Babylonian Talmud (3b) where were learn that a mystical breeze played upon the strings of King David's harp that was suspended above the King's bed, so that David arose and played the harp until the sparkling of the light of dawn:

מהנשף ועד הנשף או מהערב ועד הערב! - אלא אמר רבא: תרי נשפי הוו, נשף ליליא ואתי יממא, נשף יממא ואתי ליליא. ודוד מי הוה ידע פלגא דליליא אימת? השתא משה רבינו לא הוה ידע, דכתיב זכחצות הלילה אני יוצא בתוך מצרים, מאי כחצות? אילימא דאמר ליה קודשא בריך הוא כחצות - מי איכא ספיקא קמי שמיא? אלא דאמר ליה (למחר) בחצות (כי השתא), ואתא איהו ואמר: כחצות, אלמא מספקא ליה - ודוד הוה ידע- ? דוד סימנא הוה ליה, דאמר רב אחא בר ביזנא אמר רבי שמעון חסידא: כנור היה תלוי למעלה ממטתו של דוד, וכיון שהגיע חצות לילה בא רוח צפונית ונושבת בו ומנגן מאליו, מיד היה עומד ועוסק בתורה עד שעלה עמוד השחר. כיון שעלה עמוד השחר נכנסו חכמי ישראל אצלו, אמרו לו: אדונינו המלך, עמך ישראל צריכין פרנסה. אמר להם: לכו והתפרנסו זה מזה. אמרו לו: אין הקומץ משביע את הארי ואין הבור מתמלא מחוליתו. אמר להם: לכו ופשטו ...

However the heavenly court honored the Rabbi Teitelbaum with an angelic sermon (*derasha*) that continues in the heavenly halls until the redeemer "*mashiach ben David*" comes.

We learn from esoteric Jewish texts that the Jewish souls (*neshamah, ruah, chayah, and nefesh*) are defined by musical ratios represented on stringed instruments, and indeed the transmigration of the soul (*gilgulim*) is sent on its journey to a song. Rabbi Yehudah HaLevy poetically and philosophically wrote in a poem that one turn one's life into a song, and it is through music that all transcendence is possible. Rav Yehudah HALevy famously wrote that his pen was his harp and his library his gardens. *The Spanish poet and physician, Rabbi Yehudah HaLevy wrote, ""My pen is my harp and my lyre; my library is my garden and my orchard. 3"1 The original line is: נבלי וכנורי בפי עטי / גני ופרדס ספריה My harp and my viol are in my pen / its books are my garden and orchard of delight" This is from the poem beginning " יונה תקנ " -- that's poem #110 in volume 1 of Brody's edition (Diyan : ye-hu sefer kolel kol shire Yehudah ha-Levi ... 'im hagahot u-ve'urim ye-'im mavo me-et Ḥayim Brody. Berlin : bi-defus Tsevi Hirsh b.R. Yitsḥak Iṭtskoyski, 1896-1930). See p. 166, line 37-8*

While Kohelet might note, there is a time to sing and time not to sing, the *piyutim* (liturgical poems) that adorn and pepper the Rosh Hashanah and Yom Kippur Mahzor, represent an apex in music. It is with the Piyyuṭim that music found scope for development within the walls of the synagogue (comp. Zunz, *l.c.* pp. 7, 8, 59, 60) For example the various angels of various hierarchical ranks, in fact intoning celestial sublime song, which is not just fitting praise of Hashem who is *leilah leilah* (beyond all praise) but because of Hashem, who created the spirit of the soul, which can only find its ultimate expression in song and music.

Lag B'omer and Rabbi Bar Yochai Musical Mantra

During *Sefirah* out of deference for the disciples of Rabbi Akiva who passed on because they did not show respect for each other, instrumental music is not played but some Jews will listen to vocal Jewish music. It is on *Lag b'omer* that this ban is lifted, as the Rashbi is said to have ascended to heaven, as if on a mystical song to unite with the *Shekhinah*. It is indeed on that date of the 33rd of the Omer that the Bar Yochai *mantra* is sung with great mystical fervor. It is worthy to recall its lyrics with much kabbalistic content, and they are sung with accompanying spiritual power. Abraham Sutton translates the text and puts this in musical format of a refrain calling the text a musical (hypnotic) hymn. We read:

Rabbi Shimon Lavia on a chabad site at http://www.chabad.org/kabbalah/article_cdo/aid/379964/jewish/Bar-Yochai-Song.htm writes, "This hymn extols the virtues of Rabbi Shimon bar Yochai, the author of the holy Zohar. It relates how he achieved greatness in each of the 10 *sefirot*. Each stanza corresponds to a different one of the *sefirot* (indicated here by the Hebrew small words that precede each stanza (*malchut, yesod,* etc.). The composer spelled out his name in the first letters of the Hebrew stanzas. This song, commonly sung on Shabbat and all through the days of the counting of the Omer, is heard around the clock on *Lag B'Omer* in the northern Israeli town of Meron, the site of Rabbi Shimon's tomb.

> Refrain:
> Bar Yochai - fortunate are you, anointed with joyous oil [i.e. wisdom], over and above your companions.
>
> (*malchut*):
> Bar Yochai...You were anointed with the holy oil that flows down from the transcendent [source of mercy]. [Like the High Priest], you wore a holy crown that set you aside from other men, an aura of splendor bound eternally upon your head. (Refrain)
>
> (*yesod*):
> Bar Yochai...It was a comely dwelling that you found, on the day you ran away and escaped from the Romans. [For thirteen years] you stood in the sand of the rocky cave - there you merited to your crown of splendor and radiance. (Refrain)

Secrets of the Torah whose fragrances are sweeter than blossoms and flowers....

(*netzach/hod*):
Bar Yochai...Your students are like the [strong and beautiful] beams of acacia wood [used to hold up the Tabernacle]. When they learn Gd's Torah, they become ignited with the wondrous burning light [of its secrets]. Behold, these secrets were revealed to you by your teachers [Moses and Elijah]. (Refrain)

(*tiferet*):
Bar Yochai...[While still alive] you ascended to the Field of Apples [Garden of Eden] to gather remedies [for the souls of your people]. Secrets of the Torah whose fragrances are sweeter than blossoms and flowers. For you alone the entire creation of Man was worthwhile. (Refrain)

(*gevura*):
Bar Yochai...You girded yourself with strength and attained total self-mastery in order to fight the battle of the Torah of [black fire on white] fire in the gates [where the judges sat]. You unsheathed its sword and brandished it against the enemies [of your people]. (Refrain)

(*chesed*):
Bar Yochai...You ascended to a palace of pure light marble stones. Even there you [hardened your face like the lion, and] stood unmoved before the constellation of Leo. Crowned in glory, you ascended beyond the Great Bear [to perceive wonders that no mortal ever grasped]. You saw, but who could see you?! (Refrain)

(*bina*):
Bar Yochai …When you reached the Holy of Holies [of the Supernal Tabernacle, you grasped the secret of] the Green Line (the thread of measured light through which Gd created the world and) through which He continually renews the works of Creation daily. [The works of Creation are known as] the Seven Weeks (Forty-Nine Gates of Understanding). In order to go beyond this and grasp the secret of Fifty, you bound [your thought to] the letter *shin* [on both sides of the Head-*Tefilin*]. (Refrain)

(*chochma*):
Bar Yochai…You perceived the inner radiance of the letter *yud*, the ineffable wisdom of the Torah that preceded Creation. [You mastered] the Thirty-Two Paths [that flow from the *yud*, the essence of the Torah which is called] the "First Teruma". Then, like the Cherubim [on high], you were anointed with the splendor of [Gd's] radiant light. (Refrain)

(*keter*):
Bar Yochai…When you reached the highest level of the mysterious hidden light, you feared to gaze due to the enormity of its radiance. It [is the most hidden level of Gd's Will and Purpose which] is called No-thing, concerning which [Gd] said, "No man can see Me [and remain physically alive]." (Refrain)

Bar Yochai…Fortunate is the mother who bore you, fortunate is the nation that imbibes your teachings! And fortunate are those who grasp the secrets [you revealed]! They don the Breastplate of your perfections and lights. (Refrain)

This mystical hymn expresses all the power that music can convey and it is through this mystical hymn that we should merit to experience the true power of music which is the most holy crown of all the fine arts.

This hymn encapsulates the Rashbi's life in succinct musical form.

The oral teachings of the text of *Sefer HaZohar*, a text that never ends, as a field of force, but partakes of an endless pleromatic emanating flow of wisdom from Eden, recounts the ironic riddling mystery of the Rasbhi's magical *hevra* who converse in the Galilee, entering into something eternal. This eternality can also be conveyed by music, which gestures toward the *ayn sof*. Among the remarkable sections at the end of the Zohar is *Rav Metivta*, an account of a visionary soul journey by Rashbi and the Hevra, to the Garden of Eden, where the scholars discover secrets of the afterlife, and the mysteries between male and female governing the relationship between the G-d head and the Shekhinah, as a macrocosm radiated to the microcosm. Music too takes one on journeys and when played can be swept up by *ruach hakodesh*.

The account of Rashbi's death is thus a joyous occasion as a celestial voice announces his wedding celebration above. The circle of life does not end, just as the reading of the torah on *Simchat torah*, is brought full circle from the end of *Devarim* to *Bereishit*- so too the Rashbi at his *petirah* closes the circle in simchah before ascension.

The *Idra Zuta* (The Small Assembly), where Rashbi hints to his own resurrection symphony, a term to be employed in the first essay on *Mozart and Medicine*. The description of the last gathering before Rashbi's ascension, involves an intensity of revelation of mysteries of divine enigmatic being, based on the earlier gathering known as *Idra Rabba*

(The Great Assembly), but including holy words not revealed before. When the Rashbi ascends from this world to unite ecstatically with the Divine feminine-Shekhinah he prepares his *talmidim* to his wedding celebration above, confirming with *Pirke Avot* this world is a *prosdur* to *olam ha-bah*, prepare yourselves to enter the heavenly banquet hall (*metrakolin*). רבי יעקב אומר: העולם הזה דומה לפרוזדור בפני העולם הבא. התקן עמצך בפרוזדור, כדי שתכנס לטרקלין. In reality for the Rambam this banquet hall is nothing physical or tangible in *gashmius*. It is where the righteous sit with crowns on their heads enjoying the radicance (ziv) of the *Shekhina*. The crowns are directly proportional to the wisdom-understanding-knowledge gained in this world. This wisdom should include musical knowledge.

Hilchot Teshuvah, Sefer HaMadah, perek chet, Mishnah gimel (RAMBAM)

כל נפש האמורה בענין זה אינה הנשמה הצריכה לגוף אלא צורת הנפש שהיא הדעה שהשיגה מהבורא כפי כחה והשיגה הדעות הנפרדות ושאר המעשים והיא הצורה שביארנו ענינה בפרק רביעי מהלכות יסודי התורה היא הנקראת נפש בענין זה. חיים אלו לפי שאין אמהם מות שאין המות אלא ממאורעות הגוף ואין שם גוף נקראו צרור החיים שנאמר והיתה נפש אדוני צרורה בצרור החיים. וזהו השכר שאין שכר למעלה ממנו והטובה שאין אחריה טובה והיא שהתאוו לה כל הנביאים.

It is the ascent to this world of *olam ha-bah* that is perhaps conveyed as Rabbi Abraham Abulafia understood by the power of music.

Chapter One

Mozart and Medicine

at the end of the 18th Century[1]

This evening we are going to consider some of the medical events which surrounded Mozart's early life, as well as the details of his final illness.

Mozart was a musical genius, whose music and the events of his life have been extensively analyzed and whose legacy continues. However, his medical illness is only sparsely commented upon in the literature, and this medical history is open to considerable interpretation. The circumstances with regard to his passing at the age of 35 has always been puzzling, but I hope that we can shed some light upon this, and bring together some knowledge as to his underlying medical problems. We will evaluate whether Mozart received medical care appropriate for his time. If there was a PSRO agency, I wonder whether they would have judged that Mozart received too much care, or perhaps there was deficiency in some ways in his medical attention.

---

[1] Paper Delivered at Sinai Hospital in Baltimore Maryland as part of Lectureship on History of Medicine.

The record with regard to Mozart is rather complete. We know something about his almost every week, and sometimes every day of his entire life. This is possible through the many letters which he wrote, as well as those which were written to him from his father, sisters, and other people. These were assiduously saved and collected by his father musician Leopold who recognized early that the boy was a genius, and perhaps a book would eventually be written about him. As it turned out, there have been many books, whole libraries, written about Mozart, including an early rather comprehensive biography by Otto Jahn in 1882. There is also a recently published biography of Franz Niemetcheck, written in 1798, by a man who actually knew Mozart, and travel stories by Vincent and Mary Novello, English people, who in 1828, visited Saltzberg with the express purpose to interview Constanza Mozart, his wife, Anna Maria, his sister, and the younger son, Karl- all of whom were living in Saltzberg at that time. One of the best articles that appears in the medical literature is a report in the Journal of the Royal Society of Medicine by Peter Davis published in 1983. Some recent books on Mozart include but were not limited to: Hermann Abert's, W.A. Mozart, Piero Melograni's Wolfgang Amadeus Mozart.

The format of a CPC, a clinical pathological conference, may be helpful for us in evaluating Mozart's medical illness. Therefore, I would like to present to you a 35 year old musician, who presented to his physician with weakness and fever, swelling of legs and feet, and exanthem (rash), and then finally convulsions and paralysis.

The patient's past history indicates that he was born in 1756, in Salzberg, of a very fine artistic family. Like all children at that time, he was not breast fed, but was given water and a small amount of honey. Breast feeding, apparently was frowned upon at that time in the circles into which Mozart was born, and undoubtedly this may have contributed to the fact that five of his siblings died in childhood. Today breast feeding is considered very healthy: see work by 1) Miriam Wolfish, 2) Sharon Forman, 3) Mayer Gruber, 4) Sivan Berger, 5) Shnair Levin.

His father Leopold, as you know, was a violinist and recognized early that the young Mozart, age 2 ½ to 3 years of age, had considerable musical talent, and could learn pieces in a very rapid fashion. His father recognized that he was extremely gifted very early and determined to teach him not only music, but all other subjects, including mathematics, which he was very adept at, languages, which he learned also very rapidly.

At the age of six, Mozart and his sister, who was 5 years older, began traveling to other cities in Germany and Austria including Vienna. He is said to have had a sore throat, and possibly streptococcal infection in Lintz in October 1762, and then one week later in October 1762 he developed fever. Dr. Davis however in his report seems to feel that this may well have been *Erythema Nodosum*. He developed painful, tender, red modules over his extremities, which are described in Leopold Mozart's letters. Again in November 1762 and in December he developed upper respiratory symptoms, and fever with arthritis. The possibility that these illnesses at the age of six represented rheumatic fever with *erythema nodosum*, brought on by a streptococcal infection, is very suggestive.

Between the age of 8 and 11 the two Mozart children, his sister and himself, continued to travel extensively on behalf of music. He went to Paris, and he was in England for 18 months, and on returning to the Netherlands, he developed a very severe infection while in Hague in

November of 1765. His sister developed the illness initially. This was characterized by a febrile illness, with chills and fever, weight loss, a skin rash with boils, and a respiratory infection, so that Leopold in his letters, wrote:

> Over the next month it made him so wretched that he was not only unrecognizable, but had nothing left save for his tender skin and little bones. I had to take much care of his mouth. Most of the time his tongue was like dry wood and dirty, so that it had constantly to be moistened. Three times his lips lost their skin and became black and hard.

By the middle of January, he was recovering and able to walk aided with help. Both the children were sick at this time and were cared for by their parents almost continually on a day by day basis. Leopold gave the children his favorite remedies, "black and Margrave powders, Violet juice, and Tamerind water." Clearly folk remedies existed then as they do today. The children were seen by Dr. Hayman, while in the Hague, who bled the little girl with allegedly some improvement. Actually Amadeus' sister was feared to be lost and anointed by a priest on October 21, 1765. A consultant was brought in, Professor Thomas Schwenke. The illness has ben thought to be a streptococcal infection, or even typhus fever. Dr. Davis , as well as several other prior authors on this subject, however have considered that the two children had a endemic typhoid fever, because of the severe fever, toxemia, delirium, skin rash, pneumonitis, hemorrhagic excoliation of the oral mucous membranes, and the prolonged convalescence. I was especially struck by the severe mouth problems described with this illness, and remember that Typhoid Mouth Wash was still in the formulary of Johns Hopkins Hospital when I was a student there, and this was presumably one of the major therapeutic maneuvers in handling typhoid fever during the epidemics here in Baltimore in the early 20th century, when William Osler and others very carefully described the disease and advocated quarantine.

Niemetscheck who was a friend of the Mozart family, wrote in his biography of 1798 about this illness of the children:

> The family spent the summer of 1765 in Flanders, Bravant and Holland. During a dangerous illness which kept both children in bed for several months, the boy started on another six piano sonatas and then on his recovery he finished them. During his illness his creative energy persisted and since he was not allowed to leave his bed, he asked to have a board laid over his couch, on which he could write musical notes. Even while his little fingers were still covered with spots, it was difficult to prevent him from writing music. This story comes from reliable testimony.

Niemetscheck also described Mozart as being ungainly in appearance, with a small build and that this was due to "the overtaxing of his brain in his youth, and the lack of exercise in childhood." He was however born of good looking parents, and is himself said to have been a beautiful child in appearance. But from the age of six, he was permanently in a sitting posture, when he was also beginning to write at that time too. And how much this musician wrote particularly in his later years until his early death! It is well known that Mozart preferred to play and compose at night, and as his work was often urgent.

Massive creative output at night by Mozart does not necessarily suggest a manic depressive episode. While many musicians such as Robert Schuman, Tchaikovsky, Mahler, etc may have suffered from manic depression, perhaps Mozart's creative genius cannot be attributed to this mental illness. One seminar at the *Bard Music Festival*, in Annandale-on-Hudson, N.Y., examined Schumann's life and music, in a panel titled, "The Inner World of Robert Schumann: Manic Depression and the Creative Process," moderated by Kay Redfield Jamison, a professor of psychiatry at Johns Hopkins University and the author of "Touched With Fire," which chronicles the association between manic-depressive illness and the artistic temperament. Albert Rothenberg, book' *Creativity and madness : new findings and old stereotypes,* [published in 1994 by JHU Press} is one of many including Jeffrey Kottler's Divine madness : ten stories of creative struggle (Jossey-Bass, c2006)   on the subject. Perhaps, Bruckner's nervous breakdown and hospitalization, Berlioz's suicidal black depressions, Beethoven contemplated suicide, and Rachmaninoff affliction with deep depressions, and dedication of his Second Piano Concerto to his psychiatrist, may suggest the dark side of the aristic creative process? We must be on guard however to recall that King David himself feigned madness to escape his enemies in Psalm 34 when he disguised his sanity before Abimelech who drove him out. Walter Benjamin and Gershom Scholem it should be recalled also feigned madness to avoid service in WW I and spent the war in Basel Sweitzerland. Other's who pleaded madness to avoid something negative are Ezra Pound and the author of the OED. Pound had given anti-semitic broadcasts from Italy in WWII and instead of going to jail preferred to spend his internment in St. Elizabeth's outside of Washington. The book, The professor and the madman : a tale of murder, insanity, and the making of the Oxford English dictionary, describes how the author of the OED pleaded insanity in order to spend his internment in a mental hospital with a good library in order to do research, after he committed a crime of passion, shooting the man cheating with his wife in his own home. It is clear that many persons are interned as mentally ill who are not as illustrated by the Film One Flew over The Cukoo's Nest which depicts the practice of lobotomy performed on the main character Jack Nicholson, who was wild but not mentally ill. The fact is that Foucault has shown that the history of most mental institutions did not include nice libraries that Pound and the author of the OED enjoyed but rather were horrible places worse than jails, as shut down units where inmates were treated deplorably. Such poor conditions are depicted in the film Amadeus when the guilt ridden Salieri who is not mentally ill but suffers from a Kafkesque guilt complex is depicted in an insane asylum.

It can be imagined how so delicate a physique that Mozart must have had, would have been impaired by all-night creative episodes. These were probably the chief causes of his untimely death (if in fact it was not hastened unnaturally as myth would have it in the play Amadeus). From Niemetschek's biography then we have an idea that overwork in dedication to the artistic life, and his constantly sitting and writing, were significant factors in his illness of over exhaustion, sleep deprivation, lack of exercise and eventually his final illness.

When Mozart was 11 years old in October 1767 he seemed to have contracted small pox. Serious epidemics of small pox had raged in Paris and Vienna during this epoch. Naneryl, his sister wrote in one of the letters that her formerly handsome brother had become disfigured after small pox, exacting pock marks and scars over his nose.

In January 1772, while in Salzberg, following a tour of Italy, Naneryl in the letters describes her brother Amadeus as having another serious illness, in which he looked sickly and very yellow (jaundice). This illness has been considered perhaps to be viral hepatitis or even yellow fever. Plumbing was very primitive at this time and hepatitis rampant. Yellow fever also was an endemic in Italy at that time and of course viral hepatitis could also have worked in tandem.

Mozart and his mother went to Paris in the early part of 1778. There Mozart wrote several piano sonatas, which are full of charm, elegance, and brilliant technique. The sonatas belie the fact that it was in Paris at this time that his mother developed severe illness, characterized by fever, headaches, diarrhea, deafness, hoarseness, and eventually delirium and died., Mozart's mother also had a cough, while in Manheim, a few months earlier. She was bled on several occasions and given Rhubarb powder and lime. Lime actually contains quinine and was rightfully used to treat scurvy, particularly for Sailors who thereby derived the nickname "limies." The direct cause of her death is not clear. It has been considered to be typhoid fever, tuberculosis, or heart disease, as her death certificate stated in the historical record.

Mozart experienced illnesses in 1781 and 1783 but in general his health was fairly good while in Vienna until August of 1784. He became ill with fever, joint pains, abdominal pain, and vomiting. He was unable to travel to his sister's wedding because of this illness. He developed severe diaphoresis with drenching of his clothes. He wrote to his father:

> Four days running at the very same hour, I have had a fearful attack of colic, which ended each time in violent vomiting. I have therefore to be extremely careful. My doctor is Sigmond Barisani who since his arrival in Vienna, has been almost daily at my rooms. People praise him very highly.

Leopold, his father, also wrote in the letter that not only his son but a number of other people caught a rheumatic inflammatory fever, which became septic when not taken in hand at once. Mozart remained ill for several months, and the definite diagnosis of this illness has puzzled a number of observers. Renal colic, pyelonephritis, rheumatic fever etc have all been suggested. Dr Davis, in his article in the *Royal Society of Medicine*, feels that this may well have been the patient's first episode of Henoch-Schonlein disease, a renal disease characterized by abdominal pain and colic, and often arthritis. This does occur mostly in children, but can occur in adults. It is often associated with a rash, which however was not described at that time. However it is possible that this was the genesis of the patient's renal disease, which eventually may have been a factor in his death seven years later. During 1750 the patient at times showed evidence of depression, complaining of headaches and aching in his joints. Amadeus was in debt at this time, and falling out of favor with the musical public of Vienna. His wife frequently was alleged to be ill and took trips to Baden spas for the waters. His wife's illness has been described as probably thrombophlebitis and varicose ulcers, involving her legs. In the summer of 1790 Mozart wrote that:

> I am absolutely wretched today, I could not sleep all night from pain. I must have gotten overheated yesterday from walking too much, and then without knowing it, I have caught a chill. Picture yourself in my condition, ill and consumed with worries and anxieties.

During all this time however Mozart had a consistent output of music, working regularly, daily, and late into the night.

Concerning Mozart's terminal illness much has been written, and you may remember the episodes from the movie Amadeus. He had been working on the *Magic Flute* (*Die ZauberFlute*) and had been strong enough to conduct a performance of this opera. He was approached by an unknown tall, gentleman who requested a requiem. Jahn, in his biography of 1882, says that no sooner was the Magic Flute completed and performed, that Mozart set to work with relentless eagerness upon his still unfinished Requiem. I would like to quote Jahn extensively:

> His friend Joseph Von Jacquin, calling upon him one day to request him to give pianoforte lessons to a lady who was already an admirable performer on the instrument, found him at his writing table hard at work on the Requiem. Mozart readily acceded to the request, provide that he might postpone the lessons for a time for as he said, "I have a work on hand which lies very in my heart, and until that is finished I can think of nothing else."

This resurrection symphony in the film Amadeus later is represented in artistic creativity even trumping the finitude of mortality as Mozart works on the requiem and is indeed taken by horse and carriage to his own resting place, a pauper's grave, to the magnificence of the symphony with dramatic effect of G-d's "chariots of fire" (2 Kgs 2:11).

Other friends of Mozart remembered afterwards how engrossed Mozart had been in his task up to a very short time before his death. The feverish excitement with which he labored night and day despite his illness increased the indisposition which had early attacked him in Prague. That was when he was conducting Don Giovani. Even before the completion of the Magic Flute he had become subject to fainting fits, which exhausted his strength and increased his depression. It was in vain that his wife would return from Baden and sought to withdraw him from work and to induce him to seek relief from gloomy thoughts in a society of his friends. One beautiful day when they had driven to the Prater, and were sitting there quite alone, Mozart began to speak of death, and told his wife with tears in his eyes, that he was writing his requiem for himself. "I feel it too well," he continued "my end is drawing near, I must have taken poison. I cannot get his idea out of mind."

We will return to the question of poison problem on the differential diagnosis section our CPC. Horrified at this disclosure Frau Mozart sought by every possible argument to reason with which he was working on the Requiem was increasing his illness she took the score away from him, and called in a medical advisor, Dr. Closset."

Some improvement in Mozart's state of health followed and he was able to compose a cantata written by Schikander, for a Messianic festival which is K623, and was finished on November 15, 1791. The first performance was conducted by Mozart, himself. He was so pleased with the execution of his work, and with the applause it received, that his courage and pleasure at his art revived, and he was ready to believe that his idea of having taken poison was a result of his diseased imagination. He demanded the score of the Requiem from his wife, who gave it to him without any misgivings . The improvement however was a shorter duration and Mozart soon relapsed into his former state of melancholy, talked much of having been poisoned and he grew weaker and weaker. His hands and feet began to swell, and partial paralysis set in, accompanied by violent vomiting. Good old Joseph Diener, a porter, used to tell how Mozart had come to him in November 1791 looking wretched and complaining of illness. He directed him to come to his house the next morning to receive his wife's orders for their winter supply of fuel. Diener kept the appointment but was informed by the maidservant that her master had become so ill during the night, that he was obliged to fetch the doctor. The wife called him into the bedroom, where Mozart was in bed. When he heard Diener, he opened his eyes and said, almost inaudibly "not today, Joseph. We have to do with doctors and apothecaries today." On November 28, his condition was so critical, that Dr. Closset called in consultation, Dr. Sallaba, chief physician of the hospital. Mozart was confined to bed, but consciousness never left him. The idea of death was ever before his eyes, and he looked forward to it with composure. The success of the Magic Flute seemed likely to open a door of fame and fortune, and during his last days of life, he was assured of an annual subscription of 1000 florins from some of the Hungarian nobility, and of a still larger yearly sum from Amsterdam, the return for the periodical production of some few compositions for the subscribers. It was hard to leave his art just when he was put in a position to devote himself to it, unharassed by the daily pressure of poverty. Hard too to leave his wife with his two small children to an anxious and uncertain future. Sometimes these ideas overpowered him, but generally he was tranquil and resigned, and never betrayed the slightest impatience. He unwillingly allowed his canary, of which he was very fond, to be removed to the next room, that he might not be disturbed by its noise. It was afterwards carried still further out of hearing, Sophie Heidl who was his sister-in-law said, "when he was taken ill we made him night shirts which would be put on without giving him pain of turning around, and not realizing how ill he was, we made him a quilted dressing gown, should he be able to sit up." He heard with intense interest of the repletion of the *Magic Flute,* and when evening came, he used to lay his watch beside himself and follow the performance in his imagination. "Now the first set is over. Now comes the Queen of the Night." The day before his death he said to his wife,

> I should like to have heard my *Zauber Flute* once more, and began to him the bird-catcher's song in a scarcely audible voice. Kappelmeister Roser, who was sitting at his bedside went to the piano and sang the song to Mozart's evident delight. The Requiem too was constantly on his mind. While he had been at work upon it, he used to sing every number as it was finished, playing the orchestral part on the piano. The afternoon before his death, he had the score brought to his bed and himself sang the alto part. Schack as usual took the soprano Hofer, Mozart's brother-in-law, took the tenor, and Gerl, the base. They got as far as the first bars of the Lachrymose, when Mozart, with the feeling that it would never be finished, burst into a violent fit of weeping and laying the score aside. Late in the evening, Closset, the physician, arrived, having been at the theater earlier. He told Sussmayr in confidence that there was no hope, but ordered cold bandages around the head, which caused such violent shuddering, that delirium and unconsciousness came on, from which Mozart never recovered. Even in the latest fancies, he was busy with the Requiem, blowing out his cheeks to imitate the trumpets and drums. Toward midnight, he raised himself, opened his eyes wide, and lay down, his face to the wall, and seemed to fall asleep. At one o'clock he expired.

- Jahn, end of the quote.

From this description we see Mozart thinking and rehearsing music all the way up to his last moments that not only testifies to his devotion to music, but his belief in the ability of art to transcend even death itself. This is powerfully dramatized in the film Amadeus. The redemptive power of music thus trumps the sting of death. And indeed the historical record indicates that Mozart viewed working on the Requiem as his own personal resurrection symphony.

Davis in his paper indicates that the month earlier, in November 1791, he contracted a fever, probably an epidemic illness while at the lodge of the Freemasons. This was associated with a high fever and sweating. He also complained of suffering as noted in the Jahn's report, or pain on moving in bed, in that he had swollen feet and hands. He well may have had polyarthritis. There were violent episodes of vomiting at night, and later diarrhea. Dr. Closset, who was mentioned by Jahn, also called in a consultation, a Dr. Mathias Von Sallaba, a senior physician at the General hospital in Vienna. Dr. Sallaba noted an exanthem, because he diagnosed later " a heated miliary fever." However this exanthema was not noted by his wife or his sister-in-law, and there is some question as to whether the final illness was associated with an eruption. It is stated that he probably had a venous section performed, as well. The details however are lacking in some areas.

Concerning the diagnosis of the physicians who were not present, or what I will describe as the House Staff Diagnosis, Dr. Closset diagnosed *"Deposito alla testa"*, that is a deposit in his brain. Dr. Sallaba diagnosed a "Hitzieges Frieselfrieber" or heated miliary fever. In any event, neither of these diagnosis seem to give any real clue as to what the patient Mozart had. Later on a Goldener Von Lobes diagnosed the patient as having a rheumatic inflammatory fever. His medical input will be discussed in just a little while.

The differential diagnosis which we considered for Mozart's illness would include – poisoning. Most people do not take this very seriously although the drama and film Amadeus played into this popular myth. However, I have found it very interesting an puzzling. As you heard from Jahn's description of his final illness, Mozart indeed said that he thought he may have been poisoned, and at one time said, " I have a taste of death on my tongue." Mar Novello and her husband Vincent made a pilgrimage to the continent from their home in England in 1828, with the expressed effort to interview Constanza Mozart, his wife, Marie-Anna, his sister and the younger son Karl, all of whom were living at that time back in Salzberg. They interviewed Consanza and their book makes delightful reading, and is much of the original source of the later biographies....

This is a poison containing arsenic and lead and possibly Antimony, developed over 100 years before in the mid 1650s. The Roman police of 1659 are reported to have found a number of husbands conveniently murdered with this substance by their recently widowed wives. Small doses escaped detection, and the victim was reported to die only after a considerable period of time, without a trace of the poison. According to Mary Novello, in her entry of July 7, 1829 while interviewing Mozart's widow, Constanza said, "Some six months before his death, he was possessed with the idea of being poisoned.

> "I know I must die" he exclaimed. Someone had given me Auqa Toffano, and has calculated the precise time of my death, for which they have ordered a Requiem. It is for myself that I am writing this."

The remainder of the comments of the widow, taken almost verbatim by Jahn in the biography of 1882, indicating how his wife urged him to put aside the Requiem, and that it was only because he was ill that he had such thoughts of poisoning. He agreed, and worked on the Messianic Ode, and when it was such a great success, wanted to work on the Requiem again. Only a few days later he fell ill again with the idea that he may have been poisoned.

In the film Amadeus, Salieri was implicated in a possible poisoning of Mozart. The story goes that Salieri in 1823 when he was senile and in a mental institution accused himself of having poisoned Mozart. The poison rumor was so strong that Vienna at that time that Giseppi Carponi a friend of Mozart, came to Dr. Gouldener, who urged him to write a letter which would dismiss the evidence of poisoning, and so exonerate Salieri. Dr. Gouldener was the chief physician of the Hospital in Vienna, and he wrote a testament on June 10, 1824. This was 33 years after the death of Mozart. He states that Mozart had fallen ill in the late autumn of 1791 with a "rheumatic and inflammatory fever," which had also attacked a great many of the inhabitants of Vienna at that time, and that several patients had died with similar symptoms to Mozart. Dr. Gouldener had not personally attended Mozart, but he recalled discussing the case with Dr. Closset and Dr. Sallaba, who as we know did see the patient at home. He said that Closset had diagnosed as I indicated, "Deposito alla testa". He said that a few days before Mozart's death, he had met with Sallaba who said positively, Mozart is lost, and it is no longer possible to restrain the deposit. He said that Mozart died with the usual symptoms of a deposit on the brain, and that there was not the slightest suspicion of poisoning. He said that the statutory examination of the corpse did not reveal anything at all unusual." This was all according to Gouldener Von Lobes in 1824.

In 1834 Pushkin, the Russian playwright, wrote a play on Mozart and Salleri in which the poison theory was further developed. An opera was made by Rimski Korshakov at this time and more recently in our own century Igor Belza wrote a book further expanding upon the poison theory. A German physician, Dr. Dieter Kerner seems to be the most enthusiastic advocate of the poison theory. He has written a number of papers in this regard, believing that Mozart's death was due to chronic mercury poisoning, but there is very little evidence that Mozart unlike Schubert, had syphilis. There are those who feel that Mozart may have taken Quicksilver or mercury himself, in order to treat presumed syphilis. The other possibility that Belza has proposed is that Mozart was poisoned by his fellow masons by Mercuric Sublimate, because he had betrayed the lodge's secret in his opera, *The Magic Flute*. However the poison theory has been refuted by a number of investigators. It is difficult to be certain in that regard. Mercury was very commonly used medication at that time, and even Bright who later described the relationship between proteinuria and edema associated abnormal kidney at autopsy, used a considerable amount of mercury as treatment for edema, which he attributed to renal disease. Peter Schaffer, in his play and movie Amadeus, is only the last to usurp the poison theory to make vibrant drama.

Other possible causes and differential diagnosis of Mozart's illness of course, include tuberculosis or typhoid fever. Septicemia has been considered., Endocarditis could be considered, since it may well have been that he had an episode of rheumatic fever at the age of six. He may well have had a cerebral hemorrhage, in view of his mental changes and the possibility of a seizure. Mozart's mental changes and hallucinations with regard to poisoning with Aqua Toffano, may also represent early hypertensive encephalopathy.

Francis Carr, in a relatively recent book, *Mozart and Constanza*, has written a who-done-it, of the quality of a grade B movie, about Mozart's death. He believes that there was a "cover-up." Such conspiracy theories abound as of late. Carr indicates that Mozart's death came quickly, rather than after a prolonged illness. But who would want to kill him? He remembered a funeral quite inexpensive with the corpse dumped into common grave, so that an autopsy was impossible to perform for any evidence of poisoning. Carr latches onto an event which occurs two days after Mozart's death. This was a spectacular attempted killing of Madelaine Hofdeml, by her husband Franz, and then his subsequent suicide. Madelaina was pregnant at that time. Francis Carr posits that Madelaina who was a piano student of Mozart's was Mozart's secret lover, and her jealous husband Franz, poisoned Mozart, then attempted to kill his wife, and committed suicide, thereafter. Terrific drama for today's audience or TV miniseries soap opera or scandal sheet yellow journalism, but not for the factual historical record.

Most observers today believe that Mozart had some form of kidney disease. He was about at the right age to have glomerulonephritis, 35 years of age, and certainly had been exposed to streptococcal infections in the past. The swelling of the legs, arms, and hands as well as the central nervous system symptoms, which he developed terminally, would fit with hypertensive encephalopathy in a patient with end stage renal disease. Dr. Davis in his review feels that the patient had Henoch Schonlein purpura. The abdominal pain and arthralgias, which the patient experienced in August 1784, and later as well as the eruption which was suggested by the diagnosis of "heated military fever", have suggested that the patient had Henoch Schonlein purpura. If so, this would have been a number of years earlier than the first description of Schonlein in 1837, or by Henoch, who described the renal disease in 1874. If we had Mozart's kidneys we may well have had a chance to see a focal glomerulonephritis with ICA immunofluorescence, which is a common finding in patient's with Henoch Schonlein purpura, but unfortunately we do not even know where Mozart is buried, much less have an opportunity to evaluate his microscopic anatomy of the kidney.

Did Mozart receive reasonable medical care? You may note from our discussion that Mozart never went to the hospital. He was seen by physicians on several occasions both during his travels, and during his terminal illness at home. Physicians came to him, performing a house call. No physician has ever written what the findings were, except for Dr. Gouldener, who did not examine the patient, writing thirty-five years later. Nobody ever reports examining his urine, although this was a commonly used form of making diagnosis in the 16th and 17th century i.e. the so called "piss prophets", who would look at the urine and make certain diagnostic opinions from the turbidity and color of the urine. Thus in Shakespeare's plays we find the phrase, "what says the doctor of my urine." While in antiquity from at least Galen, thru the medieval ages (see Visi Tamas, *Medieval Hebrew uroscopic texts : the reception of Greek uroscopic texts in the Hebrew "Book of Remedies" attributed to Asaf" in Texts in Transit in the Medieval Mediterranean* (2016) 162-197) uroscopy or diagnosis by examining the urine was a medical practice (see introductory essay to Volume 2 on Essays in History of Nephrology), however it was not until 1812 when Wells who lived in England demonstrated a relationship between dropsy and coagulable urine.It was not until 1812 that Wells an American, who lived in England first demonstrated a relationship between dropsy and coagulable urine. This was later taken up by Dr. Richard Bright, who in 1827, first showed the relationship between coagulable urine, edema , and the pathology of kidneys at autopsy. Before that time, most edema was thought to be related to heart disease or some other factor.

Dr. Richard Bright's first case was John King, a thirty four year old male, about the age that Mozart died was the patient. While his occupation was different, he did imbibe in alcoholic spirits quite a bit as Mozart did on occasion, and you can see from Bright's very careful description of the clinical history and findings, that there was some similarity between these two patients. Bright's careful description not only of the clinical findings, but also of the pathology in the form of coagulable urine and the kidney at autopsy, clearly show the rapid progress that medical science was enjoying in only thirty five years following Mozart's death in nephrology. It is possible that if Mozart had gone to England, which was one of his options, and he was even encouraged to do this by de Ponti and others, he may have been one of Richard Bright's first patients. Instead of his kidneys being in some unknown grave, they would be residing in the crocks of the Gordon Pathology Museum in Guy's Hospital in London, where a number of Kidneys described by Dr.Bright can be seen to this day.

Hospitals in the early days of the 18th century were mainly places for patients to die. Patients would be sleeping several in a bed, and only a terminal case would be given a bed to die in alone themselves. It wasn't until the earl y 19th Century that hospitals in England and Paris became more modern and updated and were clinical tools of diagnosis and observations of illness allowed much more medical progress and benefit for the patient.

In a certain sense it probably was wise and prudent that Mozart never did go the the great Hospital in Vienna, since his demise may have been even more rapid, and certainly more painful then than the care and attention that was described by Jahn, in his biography of Mozart's last few days, whereby Mozart was devoted to thinking about his music, believing that art could transcend even death itself while he continued working on his own resurrection symphony in the form of the Requiem. This is what in aesthetics is called the religion of secular artistic redemption which Spinoza founded according to Leo Strauss who writes:

"He (Spinoza) thus showed the way toward a new religion or religiousness which was to inspire a wholly new kind of society, a new kind of Church. He became the sole father of that new Church which was to be universal in fact an not merely in claim, like other Churches, because its foundation was no longer any positive revelation. It was a Church whose rulers were not priests or pastors, but philosophers and artists and whose flock were the circles of culture and property... The new Church would transform Jews and Christians into human beings- into human beings of a certain kind: cultured human beings, human beings who because they possessed Science and Art did not need religion in addition. The new society, constituted by the aspiration common to all its members toward the True, the Good, and the Beautiful, emancipated the Jews of Germany. Spinoza became the symbol of that emancipation which was to be more than emancipation, but secular redemption. In Spinoza, a thinker and a saint who was both a Jew and a Christian and hence neither, all cultured families of the earth, it was hoped, will be blessed." See L. Strauss, Spinoza's Critique of Religion, "Preface", New York: Schocken Books, 1965 (17).]

Spinoza thus represents the founder of a new religion of secular cultural where art, philosophy, and science can transcend and trump the limits of mortality and provide not merely Boethian consolation, and achieving hermeneutic meaning in life, but more importantly given a redemptive quality to life, a place that religion in the Medieval ages held. That fact that Spinoza relegates science to the area of culture speaks very profoundly to the aspect of the Medical Humanities. History too plays a most important role in the Medical Humanities as Immanuel Kant referred to history as the Queen of the Humanities only little lower than philosophy itself although in antiquity Aristotle considered history and poetry low in the ranking of ultimate knowledge.

If I could put medical developments pre and post Mozart into some perspective, we would have early in the 18th century Herman Boerhaver, who was in Leiden in the Netherlands, and some of whose students came to Vienna. Stephen Halles was a minister in England, who in 1733, about the time that Back was writing, first measured blood pressure in the horse. Auerbrugger, in Vienna, did develop techniques for percussion of the chest, and this was about the time that Mozart was born. John Hunter in England was very active during the time Mozart was alive, and he was a surgeon who was interested in a variety of activities. You remember that Mozart may have had small pox in October 1767, and this was 30 years before Edward Jenner first used vaccination for prevention of small pox. We have heard that Dr. Davis thought that Mozart may have had *erythema nodosum* at one time, but it was not until 1808 that Robert Williams described the clinical syndrome of erythema nodosum. Lannec, in France, first uses a stethoscope in 1819, as the Paris hospitals became more organized and were performing much more in the way of medical care. The opening of Guy's Hospital in 1825 was followed very rapidly by Richard Bright, whose work we discussed briefly, but also by Addison, Hodgekins, and Gull, who described diseases which classically have their names. If Davis is correct that Mozart had Henoch Schonlein purpura, it was not until 1838 and 1874, respectively, that this syndrome was described. If the patient had hypertensive encephalopathy, it would have been hard to prove, since it was not until almost ninety years after Mozart's death that clinical measurement of the blood pressure became possible.

So what can be concluded? Mozart was born into a fine family of great musical and artistic talent. Mozart was educated by his father Leopold who saw that his son received an excellent education. Amadeus was a genius from early childhood, and his music, which he worked on devotedly even up to the last moments of his life, with great assiduousness, became the pinnacle for the classical period of music as described by Charles Rosen in his book, *The Classical Style*. Mozart worked in spite of a number of medical illnesses, which were reviewed in the past history I have given.

Mozart's terminal illness was rather brief. He was said to be well only several months earlier, and several weeks before, and had been working on the Messianic Anthem. He was working on his Requiem, and had completed the *Magic Flute*, with Masonic themes, before his death, but it was the composition of the Requiem, that he viewed as his own personal resurrection symphony that occupied his final moments.

Mozart probably had streptococcal infection a number of times during his life, and may have developed a terminal hypertensive encephalopathy, with uremia and swelling of his legs. He may have had polyarthritis, as well, and if he had a generalized skin rash, Henoch Schonlein purpura, may have been the factor.

Could medication or other treatments have saved Mozart? If Mozart suffered from kidney disease then Certainly dialysis or kidney transplant most likely would have prolonged his life. However modern treatment was in its infancy at that time, and polypharmacy with toxins such as arsenic and mercury abounded as common practices as well as blood letting. Even Dr. Richard Bright had very little to work with other than arsenic and mercury. Bright's favorite remedy for the disease that bore his own name, was the use of Calamol and Opium, standard practices of the day. Bright recommended to avoid chill as common sense. He writes: "Let flannel be worn constantly." Just like Sophi, Mozart's sister-in-law, who made a padded dressing gown, presumably to avoid the chill and keep Mozart warm. The skin in the late 19th century was an important pharmacological organ. When the kidney or some other organ was not functioning, it was felt that the vital humors had to be removed by sweating, and agents to promote this were directed toward the skin. Still in the 18th century the influence of Robert Burton's *The Anatomy of Melancholy* which describes the four humors was a common assumption of human physiology which modern science has replaced with the endocrine, skeletal, muscle, circulation, pulmonary, systems, etc. Of course thinking of the body as a machine or system was held as early as Rene Descartes, but today it is hoped that a physician take a holistic view of the whole body of a person with a soul and not focus on just one aspect of symptoms but see the forest for the trees of the whole physiology of the human being. Chilling in the 18th century was thought to prevent the release of fluids, which would imbalance the system and contribute further to the condition.

I don't feel that Mozart lacked any therapeutic maneuver that could have been available to significantly improve his condition in that day of science at the time. Mozart was probably bled on several occasions.

Let us consider a brief historical sketch of what is called bloodletting, later in history enacted by the use of leaches. Blood letting had been employed by medicine since antiquity. It is even debated as therapeutic in the Talmud (Yoma 84a). The Talmud expresses the merit of a bloodletter named Abba (Taanith 21b) who had separate rooms for men and women, but also insisted that women wear a special garment he had so that only the site of bloodletting was exposed. The mention of blood letting is also found in the secular texts as well (Oelius Auelianus, Acut. 3 chapter 4:34, p. 193). The Talmud further notes that Nebuchadnezzar chose for himself young people without blemish which the Talmud explains "there was not even a lancet puncture on their bodies" (Sanhr. 93b). Maimonides (1135-1204) recognized that routine phlebotomy is not advised and certainly after the age of fifty it is prohibited (Misheh Torah Hilchot Deot 4:18). The Talmud uses the terms lancet as *Kusulha* or perhaps scarifum. Rashi uses the old French phrase of *pointure de flieme* or a prick of the lancet. Another method blood letting was the use of cupping glasses. Known by the Talmud as *keren* or *coru* (horn). Whereas the physician in Judaism is a chakham, the bloodletter in the Talmud as his name of *umman* or *ummana* connotes an artisan sometimes called *gara*

(Kiddushin 82a; Kelim12:4; Derech eretz Zutta 10:2), which is the meaning of the Latin expression minutor. Latin gara is *minuens sanguinem*. In Syriac *minuens barbam*, the cutter or barber. The functions of the barber were served by the *sappur*. The *umman* refers to the Greek *aimon* which means shedder of blood (i.e. murderer) is not established as a clear etymology. The Mishnah commentator Rabbi Ovadia Bartenura mentions the interpretation of blacksmith. While the doctor or *rophe* from the root to heal in Hebrew could do surgical acts the blood letter sometimes was a different position. However a *rophe* who also did blood letting may be indicated by the phrase *rophe umman*.

In the 18th century Mozart's sister was bled during the epidemic when they were children. George Washington as you may know from history was bled also and it is felt that this contributed significantly to his medical demise from the quincy. If Mozart had only been able to live an additional thirty or forty years, the terminal illness may well have been better diagnosed, through the research of Richard Bright in London, and the clinical studies in the Paris Hospital on auscultation and revamping medical theory. However, we have to be content with the fact that while Mozart summed up the classical era in music, his medical illnesses reflected the then very limited diagnostic and clinical skills, as well as treatment possibilities, that were available at the end of the 19th century.

Let us wrap up with a review on the state of medical knowledge in the 17th century, and the way science and its subfield medicine work. Debate rages in the history of science between scholars like Thomas Kuhn and Raynaud. Kuhn in his book *The Structure of Scientific Revolutions* holds that sciences moves forward in advances by paradigm shifts. For instance Copernicus according to Jeremy Brown in his book, *New Heavens and New Earth* demonstrate that the shift from a Aristotelian-Petolymaic cosmic understanding to the Copernicun Heliocentric view constitutes paradigm shift in astronomy. Like wise Kuhn had previously argued that not only Copernicus, but their early manifestations in the work of Kepler, Newton, Galileo also constitute revolutionary paradigm shifts in scientific understanding as do the discoveries later of Heissenberg's uncertainty principle, Max Planck, and Einstein's relativity theory. Today string theory and Unified Field theory represent further potential paradigm shifts beyond Quantum Mechanics.

Raynaud (Univ. de Grenoble, France) on the other hand argues the "incrementalist theory" against the "radical paradigm shifts" in science that have revolutionary consequences in showing the importance of organized debates around scientific controversies that help confirm our knowledge about the world whereby science moves forward in slow, but steady movements that are the work not of individuals, but team work and collaboration.

Raynaud examines the following unique scientific conflicts within their socio-historic contexts: (1) Pasteur's germ theory versus the theory of spontaneous generation; (2) vitalism versus experimental medicine; (3) visual rays and the science of optics; and (4) the origins of Einsteinian relativism. One might add here Darwinian theory of evolution as applied to molecular level and the advances in genetic knowledge today. Raynaud's carefully chosen examples not only show pivotal moments in the history of science, but also help explain the origins and evolution of more recent scientific debates. Raynaud's incrementalist theory explains the advancement of science through scientific controversies, as opposed to Thomas Kuhn, who attributes progress to scientific revolutions and paradigm shifts, such as Copernicus's heliocentric model versus the

Aristotelian-Ptolemaic geocentric cosmology. While Kuhn sees science advancing out of crisis, Raynaud persuasively shows that understanding the role of controversy helps our understanding of the function of science and the importance of contemporary scientific debates. Raynaud offers a sober voice and clear commitment to pursuing scientific truths via scientific methods, as opposed to the politicization of science seen in the controversies surrounding Galileo and the Inquisition, and Stalinism and genetics. For Raynaud, science expresses more than the *zeitgeist* of an epoch; rather, science is focused on revealing truth.

A recent book by the Israeli philosopher of science and historian of mathematics, Menachem Fisch, at Tel Aviv University, science is also a mode of revealing truth. Fisch follows in the wake of Kuhn. Fisch defines the two opposing view of how scientific theories change as being established by Thomas Kuhn and Karl Popper. Fisch prefers to use the term introduced by Michael Friedman in Dynamics of Reason, as *"scientific framework transitions,"* instead of paradigm shifts. Fisch by turning attention to the role of ambiguity and indecision in science, offers a way to look at how scientific understandings change through morphing frames of reference. For example Tycho Brahe's planetary theory, Galileo's analysis of projectile motion, and Poincare's geometrical conventionalism, all preserved and in so doing precipitated the keen ambivalence that begot them, prompting other scientists to take a firmer stand and position. Thus the ambivalating potential of neighboring allied fields of science can propel change. Fisch marshals four examples in mathematics namely, the 19th century mathematical work of Whewell, Peacock, Herschel, and Hamilton's deliberations on the nature of science and mathematics. Fisch concludes that these examples argue against the Popperian notion that science moves forward by simple processes of trial and error. Fisch argues for example, that Peacock's *Treatise on Algebra* of 1830 that proposed splitting modern algebra into two related yet separate algebras, pertaining, as in Hamilton to very different spheres of mathematical activity—on the one hand arithematical algebra, conceived as the science of number and its relations, versus algebraica calculus, thus revealing a `wondrous convergence' of two sciences as owing to some `mysterious union' residing in the divine Mind (nous). Whewell, a Cambridge polymath insisted similarly to Hamilton that each of the inductive sciences comprised two integrated mutually cultivated yet antithetical components that led to one Truth by two paths. In short the four 19th C mathematicians cited by Fisch were onto something that would change mathematics and science forever. What they enacted in terms of scientific framework transition, could not be conceived of within the Poperian-Collingwoodian vocabularly and model of how scientific discovery occurs. Apropos of this book, Fisch's book, Creatively Undecided (Univ. of Chicago, 2017) is organized in chapter headings that echo musical structure, namely An Overature, Interlude, and an account of scientific rationality as a scientific semantic sonata! Much of the discussion of how science changes of course also owes its legacy to the philosopher of science who also was a mathematician, Ludwig Wittgenstein, whose *Tractatus-Logico Philosophicus* was a shift in the framework of epistemology, truly a groundbreaking work.

How do the findings of Dr. Richard Bright with regards to kidney disease factor into the debate between the incrementalist theory and radical paradigm shifts described by Kuhn? The subsequent articles in this book (vol 2) I hope will suggest that Dr. Richard Bright's work represents both a paradigm shift and the incrementalist theory of how scientific discoveries work. As regards Mozart we would only have been truly blessed with more sublime and beautiful music perhaps if Mozart had become a patient of Dr. Richard Bright in England, who was just making at the time major discoveries benefiting humanity in the area of nephrology and medicine in general which I hope you will see in subsequent chapters was truly a benefit to humanity. Since the goal of medicine is to prolong life, enhance the quality of life, eliminate pain, and do no harm, we can agree that this is a truly noble aim. Nonetheless we must admire Mozart's dedication to the musical art as the noblest of positions, that even in illness the creative artistic act can transcend the suffering of the body, by making culture, of which music may be the highest form, truly redemptive. Let us hope that all of us make our life into a work of art as Nietzsche urges, and as Mozart demonstrates, lead to our writing our own "resurrection symphony" a transcendent Requiem. The last scenes of the drama and film Amadeus make this apparent regardless to the factual incorrectness that Mozart was poisoned for as Mozart on his suffering death bed continues to compose the Requieum, the film fades into a dramatic enactment of Mozart's death and subsequent burial by horse drawn carriage, with a foreboding sky suggesting storm. The horses race and the Requiem plays, as if Mozart's soul through his musical art is transported on 'chariots of fire, 'as the music intones sublime melodies and verses of profound and redemptive transcendence regarding the condition of man, and the hope for art to transcend the limits of even mortality itself. True to his name Amadeus, the composer's music expresses a love of G-d as Amadeus means, "love of God", derived from Latin *amare* "to love" and *Deus* "God", who was actually born Theophilus Mozart, but preferred the Latin translation of his Greek middle name.

FINAL DIAGNOSIS:

Glomerulonephritis and with fine tuning, possibly Schonlein-Henoch syndrome, with terminal hypertensive encephalopathy. Several of the students still held out for poisoning with *Aqua Toffano*. We do not know more because a physician could not be found to discuss pathological findings in this case, sine autopsy permission was not granted.

Present Illness Sources and symptoms:

Latter half 1791

Depression, personality change, paranoid delusion, headache, blackouts, anemia, weight loss

Letters 20 November- 5 December 1791

Epidemic, duration of 15 days, fever, painful swelling of hands and feet, vomiting, diarrhea, partial paralysis, exanthema, venesection(s), terminal coma with paralysis of conjugate gaze

"House Staff Diagnosis"

"Un deposito alla testa"- Dr. Closset

"Hitziges Frieselfieber"- Dr. Sallaba

"Rheumatic inflammatory fever"- Dr. Guldener Von Lobes

Past Medical History of Mozart

Age	Date	Place	Symptoms	Diagnosis

6	14 Oct 1762	Linz	Catarrh	Streptococcal upper respiratory tract infection
	21 Oct 1762	Vienna	Fever and nodules	Erythema nodosum
	19 November, 1762	Vienna	Ailing, fatigue	Upper respiratory tract infection
	31 Dec 1762	Salzburg	Fever & polyarthritis	Rheumatic fever
8	Mid Feb. 1764	Paris	Fever, Soar throat, choking	Quinsy
	20 May, 1764	London	Ill for 10 days	Tonsilitus
9	August 1765	Lille	Bad cold and bronchitus	Quinsy (?}

	15 Nov. 1765	Hague	Serious Febrile Illness	Typhoid Fever
10	12 Nov. 1766	Munich	Fever and polyarthritis	Rheumatic fever [?}
11	Oct. 1767	Olmutz	Epidemic	Smallpox
14	30 March 1770	Florence	Bad cold	Upper respiratory tract infection
16	Jan 1772	Salzbug	Jaundice	Viral hepatitis or yellow fever{?]
22	20 Febr. 1778	Mann-heim	Temporary Indispositio n	Upper respiratory tract infection
25	10 May 1781	Vienna	Fever and malaise	Upper respiratory tract infection

27	May-June 1783	Vienna	Bad cold	Tonsilitis
28	August 1784	Vienna	Fever, joint pains, abdominal colic, vomiting	Schonlein-Henoch syndrome
31	April 1787	Vienna	Unknown: recurrence of 1784 illness?	Schonlein-Henoch syndrome
34	April August 1790	Vienna	Headache, joint pains, and malaise	Schonlein-Henoch syndrome
34			Symptoms of kidney disease	
35			Symptoms of kidney disease	

35	Passing Dec. 5, 1791	Vienna	Fever, shuddering, delirium, headache	Glomerul-lonephritis, Schonlein-Henoch syndrome, terminal hypertensive encephalopathy

Medical Developments Pre/Post Amadeus Mozart

1709 H. Boerhaevre 1608-1731 Leyden, Aphorisma/ Elemntia Chemicae

1733 Stephen Hales, Blood pressure measured in a horse

1751 L. Auerbrugger, Hospital of Holy Trinity in Vienna, immediate percussion of the chest

1756-1791 Mozart; [born January 27, 1756,]

1774 John Hunter, 1718-1783, Atlas of Pregnant Uterus, surgeon

1798 E Jenner, Vaccination for Small Pox

1808 Robert Williams, Erythema Nodoxia

1819 Laennec

1825 Opening of Guy's Hospital

1836 Richard Bright discoveries in kidney disease

1837 Schonlein, treatments for joint pain and rash

1874 Henoch, descriptions of nephritus

1876 Ritter Von Baach Clinical Measurements of the Blood Pressure

Differential Diagnosis (proposed illnesses)

(1) Poisoning theory

(2) Tuberculosis

(3) Typhus

(4) Syphilis

(5) Septicemia

(6) Bacterial Endocarditis

(7) Cerebral Hemorrhage

(8) Glomerulonephritis

(9) Henoch-schonlein purpura

Poisoning Theory

1791 Mozart proclaims, "I have taste of death on my tongue"

1823 Sallieri confesses in mental institutions that he poisoned Mozart

1825 Dr. Guldener considers poisoning

1825 Mary Novello interviews Constanza, Speculates Aqua Toffano

1830 Pushkin- writes "Mozart and Salieri"

1850 Rimsky Korsakov writes Opera "Mozart and Salieri"

1953 Igor Belza writes B grade novel on conspiracy theory of poisoning of Mozart

1955 Dr. Dieter Kerner, asserts Mercury Poisoning

1980 Peter Shaffer- play Amadeus and later film

Chapter Two

The Surgeon and the Composer

Theodor Billroth and Johannes Brahms

Their Relationship over 30 Years

"Brahms of course never read what Billroth wrote, while Billroth studied everything Brahms composed, adding pertinent commentaries."[1]

Introduction

What would bring together a surgeon, Theodor Billroth and a musician and composer, Johannes Brahms?   Well of course, music.   Both were born in northern Germany, Billroth in Bergen on the beautiful island of Rugen and Brahms in the slums (Gangeviertel) of Hamburg, Billroth born in 1829 and Brahms four later. Billroth as a child and adolescent excelled in musical education and playing the piano. Only his mother's insistence that being a doctor

---

[1] Absolon, Karel B., The Surgeon's Surgeon Theodor Billroth 1829-1894, Vol. 2, p. 119.

would be a smarter choice than being a musician led him away from music as a profession and guided him to college and the University of Greifswald, followed by medical school in Gottingen and surgical training under Von Langenbeck in Berlin. Brahms on the other hand pursued a musical career alone, playing dance music on the piano at age of nineteen in the dives in Hamburg and eventually going on a concert tour with Eduard Remenyi, a Hungarian violinist that brought him in contact with the leading composers of the day, and initiated his composing career. Brahms and Billroth met together for the first time in 1865 in Zurich, Switzerland where Billroth had been installed as Professor of Surgery. They met and played together. Shortly thereafter Billroth would be appointed Professor Surgery at the University of Vienna, where Brahms was living, and there began a close relationship of surgeon and aspiring musician with Brahms who by then had become the leading composer of chamber and symphony music in Europe. Billroth would play piano duets or chamber music with Brahms during these years in his large music room in his home in Vienna. The letters between Brahms and Billroth testify to their relationship and help illuminate their friendship. The first two string quartets of Brahms were dedicated to Billroth. Almost all of Brahms' music passed muster by being reviewed and played by Billroth prior to their publication. The great second piano concerto in B flat Major was sent to Billroth for evaluation in July 1881 and his returning letter to Brahms, indicates his enthusiasm and verve:

> Dear Friend: It is always a festival day for me when a manuscript of yours comes into my hands, but today it has given me special pleasure. Now at last we have the so long wished-for Second Piano Concerto. What a magnificent piece of music! It flows along without any effort; what noble and

musical music ! A happy, satisfied atmosphere envelops the whole.[2]

Billroth's Contribution to Medicine

Almost all medical students and doctors recognize Theodor Billroth as a surgeon who is associated with Billroth's operation (removal of the lower portion of the stomach (pylorus) with end- to- end anastomosis with duodenum) as well as Billroth's operation (Gastrojejunal anastomosis with duodenal closure), but few can tell you anything else about Billroth. His major textbooks and areas of interest include:

1. A classic textbook Lectures on General Surgical Pathology and Therapeutics (Die Allgemeine chirurgische Pathologie and Therapie) published in 1863 with multiple editions and translations.

2. Use of statistics to analyze results of surgery. "It was the Americans, specifically the Bureau of Statistics under George A. Otis who untiringly compiled such data from the War of Secession... This is how I got the idea to analyze my whole clinical material, in the same way the Americans did with theirs from the army hospitals."[3]

---

[2] Barkan, Hans, Johannes Brahms and Theodor Billroth Letters from a Musical Friendship.p. 103-105. The editor wrote a footnote:"Billroth wote on the evening of the day that he received this piano concerto. His musical taste, and his ability to form a concept of a new composition so quickly and to express what he does in this letter after a short acquaintance with the composition is astonishing"

[3] Absolon, Karel, The Surgeon's Surgeon – Theodor Billroth, Vol 1 ,p 275. The massive six volume set "A report of surgical cases treated in the army of the United States from 1865 to 1871" must have been available for Billroth to have reviewed.

While not as voluminous Billroth's data uses the same plan as the Americans in documenting the number of cases and the results of surgery.

3. Teaching and Learning the Medical Sciences (the German title was Lehren Und Lernen Der Medicinischen Wissenschaften) published in 1876 while Billroth was in Vienna. This book reviewed the history of medical education from ancient times to the present. He analyzed critically pre medical school training, as well as the history of medicine as an introduction to clinical medicine, critically discussing the teaching of medicine in the German Universities. Abraham Flexner who wrote a somewhat similar analysis of American medicine in the first part of the 20th century for the Carnegie Foundation, wanted to see Billroth's book published in America and induced William Welch to provide an introduction. Welch had been noted to have a reputation of being tardy in coming up with such material but as the book was translated and ready for publication, Flexner promised Welch a million and a half dollars to create the Wilmer Eye Institute. This was enough to encourage Welch to come up with the Introduction the very next day! "The money was Billroth's contribution to the Johns Hopkins Medical School"[4]

4. In 1858 while still in Berlin, completing his surgical training under Langenbeck in Berlin, Billroth published Historical Studies on the Nature and Treatment of Gunshot Wounds from

---

[4] Ibid Vol. 11 p. 178.

the Fifteenth Century to the Present Time. In 1870, having recently arrived in Vienna as Chief of Surgery he participated at the front in the Franco-Prussian War. Johannes Brahms, while not participating in the war, wrote in 1871, Triumphlied (Song of Triumph), a massive forty minute composition for orchestra, chorus, soloist in remembrance of the victory.

5. Billroth practiced surgery just before the Germ theory and practice of asepsis brought on by studies of Koch and Pasteur. He had however written a monograph on the Coccobacteria, (Streptococci or Staphylococci in chains or clumps) in 1874 but failed to accept completely the antiseptic technique of Lister using carbolic acid. "The importance of his discovery of the Streptococcus on the development of bacteriology was credited by Robert Koch in a letter to Billroth dated April 29, 1890, "As I started my first investigations, I was entirely under the influence of your studies on Coccobacteria Sepica etc."[5]

6. Pictures of Billroth operating show no sterile technique or use of sterile gowns, masks or sterile rubber gloves, or use of sterile silk sutures, the latter two developed by William Halsted, who visited Billroth in Vienna, primarily carrying away insight on the organization and training of a department of surgery which he put into practice at the Johns Hopkins Department of Surgery.

---

[5] Absolon, Karel B.,The surgical school of Theodor Billroth, Surgical, Oct. 1961, p.699.

7. Billroth used animals and cadavers to help develop new operations such as the first to resect the esophagus for cancer, larynx, stomach and portions of the intestine.[6]

Theodor Billroth and Johannes Brahms – a Musical Partnership Using Letters Between Them toCharacterize Their Relationship

"Brahms was not keen to express anything of real importance in words and the invention of the postcard was a great blessing to him. The purpose of Brahms' letters was to communicate. Billroth's letter-writing was a literary exercise, just as his medical writing was communicating, substance and style."[7]

The use of the computer has made letter writing an inadequate and outmoded device to study the interaction or in depth characterization of a subject. The brief hurried computer messages currently being generated no longer serve to detail and characterize the relationships between two individuals. The letters between Claire Schumann and Brahms or letters that form the substance of Goethe's The Sorrows of Young Werther were models that provided open vistas for the characterization of relationships and events. The letters, a total of 331, between Johannes Brahms and Theodor Billroth, Letters from a Musical Friendship from November 1865 through January 1894 as

---

[6] Rhoads, C. P., Biographical Sketch of Theodor Billroth and translator of Historical Studies on the Nature and Treatment of Gunshot Wounds from the Fifteenth Century to the Present Time, p. XVIII.

[7] Absolon, Karel B. the Surgeon's surgeon Theodor Billroth 1829-1894,volume 2 p. 195-196

translated and edited by Hans Barkan,[8] provide a starting point to characterize the personalities, character and relationships of these two individuals.

On First Playing a Brahms Sextette

The following is an early letter from Billroth, who was then Professor of Surgery in Zurich, to Brahms dated 4 May 1866. Billroth had an extensive musical education as a youngster and adolescent playing piano, violin and viola and had only recently met Brahms in Zurich. His enthusiasm, excitement and musicianship jump off the page as he describes his first playing of Brahms recently composed Sextette:

> Dear Brahms! Yesterday we played your new sextette at my home, partly with professions, partly with amateurs, and I wish to tell you what an extraordinary joy we had in playing of it. Playing it as a four-handed arrangement for piano, I could not have any realization of the extraordinarily beneficent and happy feeling. This is due not only to the ease with which the stream of melody flows and in which one charming motif after the other associates itself, but also to the entire construction of this work of art, to the crescendo of the emotions and the harmonic entity of the whole. The close of the adagio is of almost spiritual and of magical melodic feeling. etc. Yours, Th. Billroth

---

[8] Johannes Brahms and Theodor Billroth – Letters from a Musical Friendship, translated and Edited by Hans Barkan, Norman University of Oklahoma Press, 1957.

Early Letter by Brahms in Contrast to Billroth's style

The following letter from Brahms to Billroth in 1870 when both of them had settled in Vienna, indicates the compactness and lack of bombast often present in Billroth's letters, dated 1870. This was probably on a post card:

> Most Honored Herr Doktor: Could you perhaps take pleasure in and have time to try with me my four-handed arrangement of my G minor quartette? I would like to see you about that tomorrow afternoon. Your devoted, J. Brahms.

Brahms would later dedicated his first two string quartettes to Billroth and play duets of this and compositions with him frequently .

Enthusiastic Letter From Billroth on recently composed Triumphlied

The following is a letter dated Vienna, 6 March 1872 from Billroth to Brahms concerning his newly composed Triumphlied (Triumph Song). This is a large scale work for massive chorus, soloists, and orchestra not very frequently played in America. It celebrates the victory of Germany in 1870 over the French in the Franco Prussian War, a piece not well known or played in this country. Brahms traveled and wrote many works including this work following the war while Billroth volunteered as a surgeon on the front lines at a Hospital and later wrote of his wartime medical experiences in Chirurgische Briefe (Surgical Letters), description of wartime injuries and wounds and surgical management:

Vienna, 6 March 1872

Dear Brahms ! My best thanks for sending the Triumph Song, Opus 55, which gave me an extraordinarily magnificent and elevating impression. You show again a new side of your genius, a glistening mass of color and monumental musical, a musical triumphal arch of antique structure but still pleasant to the modern ear, etc. Yours, Theodor Billroth.

Billroth's surgical letters from the war hospitals in Weissenburg and Mannheim (1870) had appeared the Berlin Clinical Weekly and later in book form. Neither Billroth nor Brahms hated the French but both rejoiced at the formation of the German Empire and honored its founder, Bismarck.[9]

Continuing with the exchange of letters of Brahms and Billroth concerning the Triumph Song, Opus 55

To Billroth   March 1872   (on a visiting card)

I can't resist sending you the middle part of it (Triumph Song) also, but I must beg to have it back as soon as possible. Hearty greetings." (No signature)

To Brahms   Vienna, 12 March 1872

Dear Brahms! Best thanks for sending the middle part of your work. The human grows with aims; that also appertains here. What I especially enjoy is

[9] Barkan, Hans (Translator and Edited by)   Johannes Brahms and Theodor Billroth – Letters from a Musical Friendship,   Norman University   of Oklahoma Press, p15, 1915.

that all is so full of substance and short on form. The isolated parts of the chorus are not too long, and still every single bit of it, just as the whole, is broad and great. I would like to say that that is the modern concession to our minimal practice in hearing polyphonic music for longer periods. According to my feelings, it is, psychologically, entirely correct. Joyous movements and a triumphant mood do not need reflective and spiritual spinning out of the individual musical thoughts. As far as my talents reach, you give me, in imagination, the whole. I expect extraordinary things in its effect. You are a master of the beautiful in musical form, and you know how to fill the forms so that everyone who is earnestly interested in this art must be thankful to you and call you that which cannot be described; here it is being done. The immortal beauty attract us. Yours, Th. Billroth

Now that doesn't sound like a blood and guts surgeon! It is over the top, enthusiasm gushing forth, but with a knowledge of music and emphasizing the emotional and psychological aspect of the music – all from studying the score! This contrasts with the blunt short but practical writing of Brahms who in a short note on a visiting card wants the score back.

Brahms and Billroth on Trips Together

Brahms accompanied by Billroth took many trips together. Brahms' first trip to Italy was in April 1878, Billroth had been there many times previously. From Rome, Billroth wrote to his wife:

Brahms is entirely dissolved in admiration and joy. He is a splendid companion. I am often astonished how well prepared Brahms is for Italy, especially in the arts and historical and cultural affairs. He has the desire, as everyone has who first comes to Italy, to become familiar with everything Italian.

In another letter by Billroth, "Brahms is an excellent companion for a trip—so full of warm sensitivity for all beauty and in good humor. We do almost everything on foot and find ourselves in splendid shape. We are a peculiar three leafed clover; on the streets we are seldom together. Brahms, the youngest (by four years), always ahead, always, jolly looking into all the stores, and amused about everything; ten or fifteen paces after him come I, somewhat more slowly. We meet on corners and discuss the map of the city. Brahms bubbles with desire to speak Italian, has studied the grammar for months and learned all the irregular verb; however, he seldom finds just what he needs for the moment."

What Drew these Two Men Together, one a Surgeon the other a Musician

Brahms is reported to have said, " Billroth was attracted to my music at a time when most people did not wish to hear any of it; this friendship has been a gift of fortune, and his warm enthusiasm has become a necessity to me."[10] "Billroth, beset by the problems of life and death of the day, needed the musical joy flowing toward him from his friend Brahms, often presented in his own home by musical friends, often enjoyed by himself only in the late

---

[10]Johannes Brahms and Theodor Billroth: Letters of a Musical Friendship, Translated and Edited by Hans Barkan, Norman University of Oklahoma Press p. 247-248

evenings when a manuscript from Brahms had arrived. Billroth's desire for a musical career from childhood and adolescence, cut off by his mother and others for a career in medicine, could be experienced vicariously by his close contact with Brahms. Brahms needed the warmth and poetic extroversion of his friend; he needed, as so many essentially introverted men need, the praise of the essentially extroverted man... Both men can look down with equal satisfaction to their friendship on earth, and to that which each man meant to the other's happiness."[11]

Billroth, an experienced traveler who possessed an exhaustive acquaintance with fine art, enjoyed accompanying Brahms on many trips including his first encounter with the art and architecture of Italy. While Brahms did not enjoy Italian music, nor does any of it appear in his music, with Billroth as a guide he enjoyed Rome, Naples and Florence. The reserved Johannes, used to short pithy letters wrote, Robert Schumann's widowed wife, Clara, the following enthusiastic letter:

> How often do I not think of you, and wish that your eyes and heart might know the delight which the eye and heart experiences here. If you stood for only one hour in front of the façade of the Cathedral of Siena, you would be overjoyed and would agree that this alone made the journey worthwhile. And, on entering, you would find at your feet, and throughout the church, no single corner that did not give you the same delight etc.[12]

---

[11] Ibid, p. xxi.

[12] Geiringer, Karl, Brahms, His Life and Work, Oxford University Press, 1947, p. 135-136.

Brahms enjoyed the accompaniment and guidance of Billroth on these trips to Italy and other places, further cementing their mutual dependence and friendship.

Billroth's Illness and the Return to Music with the Writing of " Wer ist Musikalisch" (Who is Musical)

In 1888 Billroth developed pneumonia and following that a heart condition that limited his hectic surgical activities and he returned primarily to interest in music. He had read the physician and physicist Hermann Helmholtz's comprehensive book published in 1863, "Sensations of Tone", a treatise bridging physiological acoustics and esthetics. Helmholtz was also the inventor of the ophthalmoscope. He was also familiar with Edward Hanslick's much shorter volume, On the Beautiful in Music (Uber das musikalisch Schone) covering much of the same material written earlier in 1854. Hanslick was a musician and musical critic and was friend and companion of both Billroth and Brahms. In various letters to Hanslick, Billroth indicated his interest in writing a somewhat similar book. In September 1891 Billorth wrote to Hanslick from St. Gilgen, his summer home:

> About a year ago I wrote a fairly large manuscript – "Aphorisms on the anatomy and psychophysiology of musical matters." It remained untouched for a whole year; I have now taken it up again. The first chapter, "Concerning rhythm as one of the most important elementary principles of music, and one most intimately connected with the human organism," was fairly satisfactory to my critical judgment, so that I made a fair copy of it. The beginning of the second chapter also, "Concerning the relation of pitch to the human organism",

seemed good. But after that consideration of other kinds arose – speech, song, vowels, harmonics. I began to question the accuracy of some of my statements. This led me to read Helmholtz. I used the fourth edition which contains much new matter as compared with the first edition which I had studied years ago. A consideration of vowels, harmonics, speech, etc., bought me into the sphere of psychology, etc.[13]

Billroth had completed the first three chapters of his book, 1. On rhythm, 2. On the relation of pitch, tone and volume, 3 . Development of "tone-art" before his death in 1894. He had previously requested that his friend, Edward Hanslick, complete and edit the material with the request that "This manuscript is to be handed to my dear friend Ed. Hanslick, to be disposed of as he deems fit."[14] After Billroth's death the book was then edited by Hanslick, published using the final chapter, Wer ist Musikalisch (Who is Musical) as the title.

"Billroth was at the forefront of musicologic theory in the way in which he developed from Helmholtz his discussion of the effects of dissonance, the quality of musical notes and the origin of combination tones. In a way Billroth brought the two worlds – Helmholtz's scientific world and Brahms' musical one – together."[15] "Perhaps one of the most amazing examples of Billroth musical ability lies in his 1881 analysis of Brahms' Piano Concerto No 2 , (which

---

[13] Hemmeter, John C., Master Minds in Medicine, Chapter IX, Theodor Billroth, Musical and Surgical Philosopher, A Biography and a Review of his Work on Psycho-anatomic-Physiological Aphorisms on Music, The Johns Hopkins Hospital Bulletin, December 1900, p. 179.

[14] Ibid, p. 180.

[15] McLaren, N and Thorbeck R. Little-Known aspect of Theodor Billroth's Work: His Contribution to Musical Theory, World J. Surgery 21, p. 570.

I had quoted in the introduction of this paper). Brahms had sent his friend, the score in the morning, and by the same evening Billroth had replied – not only expressing his pleasure in the work but offering technical comments on the overall structure and on the details of the work."[16] For those who are not familiar with the Brahms second piano concerto, I would recommend as a priority, a recording of the work or a reading of the score itself, as Billroth did that day!

Summary

Billroth and Brahms had a unique musical relationship from their first meeting in Zurich in 1866 extending for the next 25 years. Billroth as a youngster and in his teen ages gave all his attention to music, gradually focusing on medicine at school and becoming one of the leading surgeons of the last half of the nineteenth century, but never giving up his interest in music. Brahms focused only on music from the very beginning and appreciated the enthusiasm, encouragement and friendship of Billroth, especially in their years together in Vienna. The letters between the two demonstrate the dependency of one to the other, Brahms who described himself as having "no opera and no wife", found in Billroth a willing admirer, absorbing almost every composition he wrote with feedback. Brahms spoke of their first meeting in Zurich and how much he valued the qualities of this great person. " Billroth", he said, "was attracted to my music at a time when most people did not wish to hear any of it; this friendship has been a gift of fortune, and his warm

---

[16] Ibid. p. 569

enthusiasm has become a necessity for me."[17] Billroth found in Brahms the vicarious reawakening of his youthful interest in music to balance his work as a surgeon.

Such was the relationship of Johannes Brahms, the composer, and the surgeon, Theodor Billroth.

---

[17] Barkan, Hans, Johannes Brahms and Theodor Billroth, Letters from a Musical Friendship, Norman University of Oklahoma Press, p247-48

Chapter Two

Chevalier John Taylor, Johann Sebastian Bach and George Frideric Handel

Did the Chevalier really operate on both Bach and Handel for Cataracts with Disastrous Results?

I recall an old wives tale that a traveling English oculist of ill repute, Chevalier John Taylor, had performed cataract operations, unsuccessfully resulting in blindness , on the two leading composers of the first half of the eighteenth century, Johann Sebastian Bach in Germany and George Frideric Handel a German, but an established English composer for many years. Was there any truth is this scenario and who was this traveling English oculist, Chevalier John Taylor?

History of Operation for Cataracts

The history of surgery for decreased vision is very ancient. Couching (Fr. Coucher-lying down) or depressing the lens is mentioned in the *Code of Hammurabi*[1] as well as by

---

[1] Jackson, David M., Bach Handel, and the Chevalier Taylor, Medical History, 12:4(1968:Oct.) p. 386.

Celsus and in ancient India, before even mentioned in the West.[2] Elliot describes two types of operations for couching: a. "a needle-like instrument, often a large hard thorn with a very sharp point is thrust suddenly through the cornea into the periphery of the lens, then using the corneal wound as a fulcrum, the operator raises his end of the instrument and thereby depresses its point lying in the lens caries the lens backward and downward."[3] The alternative approach is laterally, making an incision in the sclera depressing the lens. Elliot describes the results of couching as "deserving to be described as appalling with complications including iritis and irido-cyclitis due to the absence of sterile technique, glaucoma, missed diagnoses, etc. The ophthalmoscope was to be developed by Helmholtz not until 1850 so that evaluation of the retina and its vascular was not possible in the 18th century.

While couching for the cataract was the standard of practice in the middle of the eighteenth century Jacques Daviel in France in 1752 published the results of his work on removal of the lens, rather than couching, which marked the beginning of the modern approach to cataract surgery.[4] He considered it was dangerous to use a sharply pointed instrument thrust into the eye to perform couching. He developed a different set of instruments to actually remove the lens and avoided many of the complications of the standard therapy then current, but this required over one hundred years before it was accepted and adopted as standard therapy. Understanding the problems of infection and use of antibiotics resulted in

---

[2] Elliot, R. H. Hunterian Lectures on Indian Operation of Couching for Cataract, Lancet I p. 325, 1917.

[3] Ibid p. 325.

[4] Hubbell, Alvin, A. , Jaques Daviel and the Beginnings of the Modern Operation of Extraction of Cataract, AMA, Vol. 39, July 26, 1902,

cataract extraction being today one of the most successful types of operations.

Medical therapy in the middle of the eighteenth century consisted of bleeding, leeches and laxatives for almost all conditions considered to be inflammations and probable had about as much success as couching. R. H. Elliot in his report on the Indian operation of couching for cataract indicates that, "The testimony of many reliable witnesses has established the fact that the results obtained by the coucher deserve to be described as appalling."[5]

Chevalier John Tayor - Traveling Opthalmiater

George Coats has written the definitive biography on John Taylor.[6]

> In the eighteenth century ophthalmology had not yet vindicated in England, its position as a separate branch of practice. It was the province of a set of ambulant practitioners who toured the country accompanied by all the apparatus of shameless advertisement (including "monkeys", we are told), couching cataracts, and selling infallible salves and eye washes. This taint of quackery appears to have deterred respectable surgeons from meddling much with the subject; their operative experience was probably small and the procedure of couching, attended frequently with brilliant immediate, but disastrous after results, was likely to be performed with fewer scruples by itinerant oculists here to-

---

[5] Elliot, R.H., Indian Operation of Couching for Cataract, The Lancet March 3, 1917.

[6] Coats, George, *Royal London Ophthalmic Hospital Reports*, vol xx, May 1915 p. 1.

day and gone to-morrow, than by settled practitioners who had to abide the consequences of their handiwork."

Coats points out that the designation of Chevalier before his name was a self appointed title. In his autobiography, The History of the Travels and Adventures of the Chevalier John Taylor, *Ophthalmiater* (a word he made up from the Greek- eye and physician). He was born in 1703 and studied medicine at St. Thomas's Hospital in London and was one of few surgeons to practice eye surgery with a medical degree. He studied under William Cheselden who had an interest in eye surgery as well. The title page of his Travels and Adventures contains the Latin motto, Qui visum vitam dat"-who gives sight gives life.

Taylor describes in his Travels and Adventures in self laudatory fashion the extent of his considerable travels:[7]

I set our form my native country and began my travels in the year, 1727

I was in progress trough every town in all England, without exception, tot the end of the year 1728

In this month I went to Paris and after a few months being there, I went through all France, every town of any consideration without exception; and thence thro' all Holland, and every town, without exception; and all this with such amazing rapidity that I was returned to London in November 1735

Being at that time called, though in depth of winter, the court of Mecklenburg, for the recovery of the

---

[7] Taylor, John Chevalier, Ophthalmiater, The History of the Travels and Adventures of the Chevalier John Taylor, Opthgalmiater. London: Printed for J. Williams, on Lidgate Hill, 1761 p.5-14.

sight of the then reigning prince; and having restored the sight of that prince, I left that court in the middle of March, in the same year, and proceed for Hamburgh and Denmark, whither I was called; I arrived at the court of Copenhagen about the middle of April, 1751.

Coats describes the Chevalier's entrance into a town traveling with "no less than two coaches and six, above ten servants in livery...It is said that his coach was painted over with eyes, and bore the motto which occupies the title page of nearly all his works, Qui visum vitam dat. He demonstrated to the spectators who viewed his entry into town his "magnificent array of instruments". He then would give a lecture on eye diseases and their treatment as well as anatomy of the eye. "His arrival was accompanied by showers of leaflets, placards, and advertisements."[8]

Chevalier Taylor practiced the standard procedure of the day of couching for cataracts which had been previously described by Petit – opening the capsule, making a hole in the vitreous and pushing the cataract into it. Taylor described this procedure in one of his early writings from 1736, *A New Treatise on the Diseases of the Chrystalline Humour of a Human Eye*. The patient sat in a chair facing the operator with an assistant behind the patient holding his head firm and steady since no anesthesia was given. Post operative care might consist of Peruvian Balsam, and warm water which was dropped in the eye, then a cataplasm (a poultice or soft external application) with pulp of cassia. On the second day a spirituous fomentations (like a poultice) with camphire and gentle

---

[8] Ibid. p. 14.

evacuations were continued for twenty days.[9] A firm tight bandage was also applied for a week, as well, hastening possible infection.   There was often immediate improvement in the patients vision since the offending cataract had been displaced. The itinerant Chevalier Taylor would pronounce a favorable result and be off to a neighboring town, oblivious of the follow up report, before the often disastrous results were apparent.

Coats summarize the Chevalier Taylor.   "A striking and picturesque figure, of good person and address seemingly, with something not unattractive in his manner; at any rate wholly unembarrassed by any diffidence in his dealings with the great. An unparalleled liar, pre-eminent among charlatans in the arts of advertisement; with a natural aptitude in the grand style...His showy qualities undoubtedly imposed upon many intelligent contemporaries, but the more discerning he seems to have been regarded as an amusing rascal...Many elements go to the formation of the complete charlatan –bombast, effrontery, dishonesty, ignorance."[10]

Did the Evidence Support that Chevalier Taylor Operated on Both Bach and Handel?

In his book, Travels and Adventures Taylor gives evidence to the fact that he operated on both Bach and Handel:[11]

---

[9] Jackson, David M., Bach, Handel and the Chevalier Taylor, Medical History, 12:4 ()ct-1968) p. 387.

[10] Taylor, John Chevalier, The Travels of the Chevalier John Taylor. p.61.

[11] Ibid. p. 25.

But to proceed, I have seen a vast variety of singular animals, such as dromedaries, camels, and etc. and particularly at Leipsick, where a celebrated master of music, who had already arrive'd at his 88th year, received his sight by my hands; it is with this very man that the famous Handel was first educated, and with whom I once thought to have had the same success, having all circumstance in his favour, motions of the pupil, light, and but upon drawing the curtain, we found the bottom defective, from a paralytic disorder.

Coats correctly points out that "these statement contain a good average of inaccuracies". 1. Bach never lived to age 88 but died at 65 and the operation was a failure and he never "received his sight" but became blind. 2. Handel was not "first educated by Bach" , in fact while the two were born in Germany they never met and Handel's music education was from other sources . 3. Handel was couched in 1752 by William Bromfield, a surgeon to St. George's and Lock Hospital and although the operation wasn't successful he did not become completely blind.[12]

The only other evidence that Taylor operated on Handel is contained in an anonymous poem from the London Chronicle, 24th August 1758:

On the Recovery of the Sight of the Celebrated Mr. Handel by the Chevalier Taylor

From the hill of Parnassus adjourning in state,

On its rival, Mount Pleasant, the Muses were sate;

---

[12] Coats, George, The Chevalier Taylor p.7.

When Euterpe, soft pity inciting her breast,

Ere the Concert begun, thus Apollo address'd:

Great Father of Music and every Science,

In all our distresses, on thee our reliance;

Know then in you villa, from pleasures confin'd

Lies our favourite, Handel, afflicted and blind.

For him who hath travers'd the cycle of sound,

And spread thy harmonious strains the world round,

Thy son Aesculapius' art we emplore,

The blessing of sight with a touch to restore.

Strait Apollo replied: "He already is there;

By mortal's call'd, Taylor, and dubb'd Chevalier:

Who to Handel (and thousands beside him) shall give

All the blessings that sight in old age can receive."

While Taylor did operate on Bach on two occasions as will be seen, we have only these two pieces of questionable evidence, the statement in Taylor's, Travels and Adventures and the anonymous poem quoted above to support the evidence that the Chevalier Taylor operated on both Bach and Handel.

Encounter of Bach and Chevalier Taylor

Bach is thus a terminal point. Nothing comes from him; everything merely leads up to him to give his true biography is to exhibit the nature and the unfolding of German art, that comes to completion in him and is exhausted in him, -to comprehend it in all its strivings and its failure. This genius was not an individual, but a collective soul. Centuries and generations have labored at this work, before the grandeur of which we halt in veneration. To anyone who has gone through the history of his epoch and knows what the end of it was, it is the history of that culminating spirit, reaching its apex in a single personality.

Albert Schweitzer, J.S. Bach, Vol 1, *The Roots of Bach's Art*, p. 3-4.

The exact nature of Bach's visual problems is in doubt. The standard portrait of Bach by Haussmann shows his eyes to have a narrow palpebral fissure suggesting to some that he may have been myopic . Others considered his use of his eyes in long hours copying manuscripts and writing music as factors. In his old age he developed a painful eye disease. It was suggested that, "his age, the sudden onset of violent pain in the eye, the lack of perception in the last stages, and finally the stroke preceding his death –points to glaucoma, possibly hemorrhagic."[13]

At the urging of his friends he became a patient of Chevalier Taylor who had arrived in Leipzig with his usual entourage and the following report appeared in the press on April 1, 1750:

---

[13] Snyder, Charles, Archives of Ophthalmology, 69, 1951, p. 832.

Among others he (Taylor) has operated upon Capellmeister Bach who, by constant use of his eyes had almost entirely deprived himself of their sight, and that with every success that could have been desired, so that he has recovered the full sharpness of his sight, an unspeakable piece of good fortune, which many thousand of people will be very far from begrudging this world--famous composer and for which they cannot sufficiently thank Dr. Taylor"[14]

This type of press release could have been under the direction of Chevalier Taylor's front men. In any event any possible relief was short lived and a second operation was performed after several days without success, after application to the eye of "local irritations of the eye by repeated incisions and cataplasms with excessive use of the entire dubious armamentarium of the times including calomel, cantharides, bleeding, etc.[15]

However Bach did not respond and became blind and terminally had a stroke and on July 28, 1750 . Taylor as was his habit left Leipzig shortly after his procedures for Berlin and on April 23 1750, notice appeared in the newspaper, the Spenersche Zeitung:

The well-known oculist, Dr. Taylor, who arrived here a few days ago from Potsdam, was obliged to leave here again last Monday. The reason for this abscondence is no other than that, Taylor having operated upon two blind women at Potsdam, the results turned out to be so bad, that both women suffered infinitely from this operation and, to all appearances, will lose their eye sight completely

---

[14] Ibid. p. 832

[15] Ibid. p. 832

whereuponHis Gracious Majesty most graciously issued the order for the said Taylor to leave Berlin the sooner the better, so that on other persons who suffer misfortune with their eyes should be exposed to such accidents as befallen those at Potsdam."[16]

While Taylor's "abscondence" from Berlin was not directly related to his unfavorable results on Bach, it was the two blind women in Potsdam that led to this order. The newspaper report of the *Leipzig Spenersche Zeitung* of July 31, 1750 seemed to place the case of his death directly to the Chevalier Taylor:

On Tuesday last, the 28th instant, died here the famous musicus, Herr Johann Sebastian Bach, Capellmeister to Their Serene Highnesses of Saxe-Weissenfels and Anhalt –Cother, Director Chori Musici and Cantor of St. Thomas's School here , in the 66 year of his like, as an unfortunate consequence of an operation very badly performed on his eyes by a well known English oculist. The loss of this extraordinarily skilled man is much deplored by all true connoisseurs of music.[17]

Bach had been working on the Art of the Fuge, his last work, but was unable to complete this fuge,which was to use the letters of his name (B A C H) and substituted a chorale, Von deinen Thron tret ich hiemit (Before thy throne, my God I stand) which was dictated by Bach a few days before his death to his son-in –law, Altniko and added to the first edition of the Art of the Fuge in place of

---

[16] Lenth, Bert, Bach and the English Oculist, Music and Letters, Vol 19, No. 2 1938, p194.

[17] Ibid. p. 196

the incomplete fuge.[18] Forkel, a biographer of Bach indicates,"of the art displayed in this coral, I will say nothing: it was so familiar to Bach, that he could exercise it even in his illness. But the expression of pious resignation, and devotion in it, have always affected me whenever I have played it; so that I can hardly say which I would rather miss-this choral, or the end of the last fugue."[19]

Handel: Visual and Neurological History and Contribution, if any, to Chevalier Taylor's Contribution to his Blindness.

Handel is universally and uniquely known as the composer of the Messiah. But it was not the Messiah that Handel first entered the ranks of the very great, though it was this work that made his name a household word.[20]

Bach and Handel were born in Germany within a month of each other in 1685 and while both represented the culmination of the Baroque musical style they never met, Bach remaining in Germany and after working as a musician in various cities settled in Leipzig. Bach was born into a family that boasted a pedigree of six generations of prior musicians so that his occupation was preordained. Handel's father was a barber-surgeon and in spite of Handel's early promise as a musician he had to rebel against his father's wishes that he pursue a law

---

[18] Jackson, David M. Bach, Handel and the Chevalier Taylor, Medical History, 12:4, Oct. 1938, p. 388.

[19] Forkel, J.N., Life of John Sebastian Bach, London, Printed for T. Boosey and Co. Holles-Street, Cavendish-Square, 1820, p. 91.

[20] Lang, Paul Henry, Gorege Frideric Handel, 1966, p5.

degree. "His musical abilities must have been noticed at an early age, but the dour surgeon paid no attention to such frivolities as music and he preferred a lawyer's career for his son."[21] Handel after a sojourn in Italy brought his version of Italian opera to the London public were it flourished for twenty five years and when it became less fashionable morphed his music into the Oratorio, showing his business and acumen, enticing the English public back into his favor. Bach on the other had initially in his career initially wrote instrumental works always with emphasis on polyphony and fugal writing and after settling in Leipzig composed weekly church corals and Passions, such as the St. Mathew Passion.

Their medical history also differed, Bach having visual problems and then probably a stroke only in the last year of his life, dying in July 1750. Handel's medical history is more complicated beginning, thirteen years earlier in 1737 when he had a stroke involving his right arm and hand with some mental changes as well. After a stay in Aix-la-Chapelle, a spa in France where he received hot water baths and other treatments he improved and the London Daily Post was able to announce in November 7, "Mr. Handel is back from Aix-la Chapelle greatly recovered in his health."[22] Again in 1743 transient motor involvement of right arm with some speech impairment was noted prompting a visit to an English spa, Turnbridge Wells. This pattern of speech and palsy was to reoccur again in 1745 with recovery.

Handel's visual problems began in February 13, 1751 he noted in German in the score of an oratoria, Jephith, which he was writing that there was reduced vision in his left

[21] Ibid, p.11.

[22] H. Bazner, M. Hennerici, Georg Friedrich Handel's Strokes, p. 152-53.

eye. No pain or other symptoms noted. ("Biss hierher komen den 13. Febr. 1751 verhindert worden wegen relaxation des gesichts meines linken auges" - "Got as far as this on Wednesday 13th February 1751, unable to go on owing to weakness of the sight in my left eye.")[23] For a while he was unable to play the harpsichord and again went to Cheltenham to take the waters in May and June, 1751 and on his return consulted Samuel Sharp, eye surgeon to Guy's hospital who had been a student of Cheselden, as was Chevalier Taylor. Sharp was later to describe a new type of knife to incise the cornea. Dr. Sharp apparently found no evidence of a cataract and diagnosed, an incipient gutta serena(impaired vision without definite cause).[24] However, the next year, November 4, 1752, there appeared in the General Advertiser, "Yesterday George Frideric Handel, Esq; was couched by William Bromfield, Esq; surgeon to Her Royal Highness the Princess of Wales, when it was thought there was all imaginable Hopes of Success by the Operation, which must give the greatest Pleasure to all Lovers of Musik, Alas no! Blind he was to remain, as his mother had been in her old days."[25] Handel however continued to conduct and play the organ and revise his works with help of a student.

Bazner and Hennericii have discussed the differential diagnosis of Handel's cerebrovascular disease possibly

[23] Hennerici, H. Bazner, M., Georg Fredrich Handel's Strokes, Neurological Disorders in Famous Artist. Front Neurosci, Basel, Karger, 2005, p. 155. (Copy of score of Jephta and Handel's notation in German from the British Museum.

[24] Hawkins, J. . History of the Science and Practice of Music, London 1853, p. 910

[25] Deutsch, O. E. , Handel: A Documentary Biography, New York , 1954, p. 726.

related to visual impairment followed by blindness.[26] In view of the recurrent palsies involving the right side with speech disturbances, the consideration of left internal carotid disease or small embolic strokes could be considered. Central retinal occlusion could explain the vision loss as well as carotid artery disease with optic neuritis. The authors suggest ischemic optic neuropathy as the most likely visual diagnosis. However without actual medical reports and only newspaper reports of the medical events it is difficult to speculate further on the neurological or visual impairments of Handel.

Finally, it has been reported that Handel and Chevalier Taylor were both in Tunbridge Wells in August, 1758. There is little evidence, however, to support Taylor operating on Handel there except for the anonymous poem published in the London Chronicle on August 24 where Euterpe (muse of the flute) calls on Apollo and Aesculapius to help the blind Handel, Apollo saying that Aesculapius is not needed because Taylor will take care of it. Likewise there is Chevalier Taylor's unreliable boast that,"(Handel), whom I once thought to have had the same success (as with Bach), having all circumstances in his favour, motions of the pupil, light, etc, but upon drawing the curtain, we found the bottom defective, from a paralytic disorder" Such a poem and braggadocio do not lend credence to Chevalier Taylor's part in Handel's blindness. Taylor however may have been right that the cause of Handel's blindness was primarily a neurological impairment.

---

[26] Bazner, H. and Hennerici, M, George Friedrich Handel's Strokes, Neurological Disorders in Famous Artists, 2005, p. 156-57.

Discussion

Bach and Handel's mastery culminated the Baroque style of music. They were followed by the classical composers, Hayden, Mozart and Beethoven in the latter part of the 18th and early 19th centuries and later the Romantic composers. What was the status of medicine in the first part of the 18th century when Bach and Handel were producing music even today envied and appreciated? Medicine in the early 18th century had made little progress from its ancient roots. Boerhaave in Leyden,during the period of Bach and Handel, was noted as a teacher and clinician, active in growing and studying plants for medicinal purposes. The name dropper, Chevalier Taylor alleged that Boerhaave, "continued me his correspondence and friendship to this latest hours."[27] Medicine had yet to achieve the prominence and credibility until it blossomed out in the 19th century in France and then in Germany. Medicine including ophthalmology was a late bloomer and lagged behind the brilliance that music had achieved in the period of the high Baroque. Couching for the cataract was a brutal and ineffective procedure lasting to the late 19th century. Only after Daviel's procedure of lens removal , Helmholtz's invention of the ophthalmoscope in 1850 making more accurate diagnose possible and surgery in the 20th century adopting sterile technique was cataract surgery a viable option. Compared to music medicine was a late bloomer. While Bach and Handel were producing incomparable music, equally appreciated today, medical care consisted of taking the waters at some spa, being bleed or having leeches placed were the principal remedies that the doctor had to offer.

---

[27] Taylor, John, The Travels and Adventures of the Chevalier John Taylor

Summary

1. This chapter evaluates the involvement of the Chevalier John Taylor in the visual problems of the musicians of the high Baroque, Bach and Handel.

2. The history of cataract surgery including the use of couching with displacement of the lens downward is discussed.

3. Chevalier John Taylor a physician and oculist is presented by his biographer, George Coats as an "ambulant practitioner with unblushing effrontery, blatant self-self-laudation and all the methods of the charlatan"

4. The evidence of the involvement Chevalier Taylor in the visual problems with Bach including the two operation s he performed for cataract and the devastating results.

5. Handel's medical history, neurological and visual are presented and while there may be some evidence of them both being together in Tunbridge Wells, where Handel was taking the waters for his condition, except for an anonymous poem suggesting that Chevalier Taylor was being called to remedy the situation and the dubious report of success in Taylor's Travels and Adventures, there is little documented evidence of his role in Handel's blindness.

6. The rudimentary status of medicine in general and specifically ophthalmology in the first half of the 18th century is compared with flowering of music under Bach and Handel in this period.

Historical Overview of Jewish Opthamologists and Treatises on Vision

by David B. Levy

The above essay makes clear that Chevalier Taylor and other itinerant quacks suggest the lowly condition of eye practice during the time of Bach and Handel.

However since antiquity Jewish physicians have shed much light on eye diseases, blindness, and opthamological complications, and we as their 'pupils' can learn much from the light of their story of the eye.

Six Jewish scholars who bring to light the Jewish contributions to Opthomology are (1) Dr. L Kotelmman of Hamburg who wrote a 426 page treatise on the anatomy, physiology, pathology, and therapy of the eye titled *Mitteilungen zur Geschichte der Medizin* VI; (2) Julius Preuss in his work *Biblische und Talmudische Medicin* devotes 28 pages to opthamology, (3) Julius Hirschberg's nine volume *Geschichte der Augenheilkunde* (1899-1918) is the most comprehensive and detailed work of its kind drawing on knowledge of Greek, Latin, Arabic, and a myriad of Romance languages; (4) Hugo Magnus published *Die Anatomie des Auges ei den Grieschen und Romern* 1878, *Die Anatomie des Auges in ihrer geschichtlichen Entwickelung*, 1900, and in 1901 *Die Augenheilkunde der Alten*; (5) August Hirsch (1807-1894); 6) Harry Friedenwald in his volume 2 of set *The Jews in Medicine*, has an excellent chapter on Ophthalmolgic Notes of Jewish Interest, (p533-550) upon which my historical overview is drawn.

In Talmud

Rabbi Muna taught, "the hand which without having been washed touches the eye deserves to be cut off, for it causes blindness (Hitschberg, Geschichte, XII, 29). The physician

Mar Samuel (165-257) known as Samuel Yarhina'ah (165 CE-257 CE) is said in Midrash Eichah (I, 16) to be learned in cures of the eye and he was the author of a remedy as well as an advocate for cold water applications

Geonim

The teacher of Rhazes (Al Razi, ca 864) who wrote over 200 medical works including a medical *Encyclopedia* (al Hawi), was a Jew and a leading physician Ali b. Sahl (Abul Hasan) b. Rabban al Tabari, the son of a noted Persian physician, and became the physician of the Caliphs I the middle of the 9th century. He is cited by Rhazes as an expert on ocular affections.

The physician philosopher Isaac Israeli (Yitchak ibn Sulaiman al-Israeli) was a leading ocular expert of Cairo before becoming a physician to the Sultan in Kairwan. He passed on around 962 CE. The *Liber de Oculis* was published in 1515 and Constantinus Africanus translated it from Arabic into Latin.

Rishonim

King Charles of Anjou, King of Sicily commissioned the Jewish translator Faraj b. Salim from Girgeni (Agigento Sicily) to translate Rhazes medical encyclopedia *Kitab al Hawi* into Latin which was finished in 1279, which contains a section on diseases of the eyes.

The Jewish Physican Nathan ha-Meati of Cento, translated Ammar ibn Ali al Musuli's work on diseases of the eye ,during his sojourn in Rome between 1279-1283 for his friend the Jewish physician Isaac ben Mordecai who was the court physician of Pope Nicholas IV and Boniface VIII.

Hirschberg also notes the translations of Zahrawi (Abu i-Qasim, Albucasis) were rendered into Hebrew, including eye surgery describing 20 operations on the eyes. This is detailed in Luisa Maria Avide Camra's *Tractado de Offalmoggia en Abulcasis*.

El Mouaffeq ben Chaoua was an Egyptian Jewish physican in Cairo who served Saladan and practiced curing disease of the eyes. He passed on in 1183.

In the *Mahaberes Ha'aruch* of the Spanish Jewish physician Solomon ibn Parhon, written in Salerno in 1160, the *sis* or swallow bird (Jeremiah 8:7) "knows and herb good for the eyes." Apparently this was discovered when a young bird swallow was blinded and put back on the nest, and the parent birds come and bring the herb placing it on the babies eyes and sight is restored." The baby bird is then sacrificed and dried and application is made of the remaining parts to those with sore eyes. Others burn the head and grind it to ashes and apply these to the eyes." This sounds awfully superstitious, and similar to the questionable practice that if one wants to see demons one should sacrifice a black cat and put its ashes on ones eyes, etc. Maimonides insisted on avoiding such superstitious practices. Rambam forbid a physician to utter an incantation over a wound and looked down upon fools who believe in the power of amulets and other magical practices. In Numbers 23:23 it states there is no enchantment with Jacob or divination with Israel".

Asad el Mahalli was a Jewish physician in Cairo in the early 13th century wrote a treatise on vision.

Maimonides (1135-1204) notes in his Aphorisms: A person who has never seen ophtalmia when he sees a case will have his eyes become filled with fluid and thus should wash his hands. If he continues to gaze upon it, he himself can be affected with opthalmia." Thus knowledge of airborne viruses.

Ibn Al Naqid of Cairo (passed on 1188-9) was the author of a treatise on the eyes.

Benvenutus Grapheus and Zacharias were two medieval opthamological authors. Hirschberg believes Grapheus was Jewish based in a Parisian codex referring to him as *"bien venu raffe"*. i.e. *ha-rophe*. Nathan b. Zacharias was a physician in Montpellier France around 1171. Montpellier in the medieval ages and after was a great medical center where many prominent Jewish physicians trained and taught. This included Jewish physican Jean Astruc (1648-1766) whose history of the medical center is a classic. There he documents the involvement of Jewish physicians such as the Saporta family, the Rabbi Yehudah ibn Tibbon (1120-1190) and his son Shlomo, and Antoine d'Aquin (ca 1648), whose grandfather was the chief Rabbi of Avignon. In the early days of the medical school Hebrew was a language of instruction along with Arabic. In the 12th century this shifted to Latin. In the 13th century Jewish physicians were expelled for some time. Lunel was another medical center in France where Jewish physicians also made a positive contribution.

Jewish physicians brought opthamological knowledge into Spain and Italy in the Late middle ages as translators and practitioners. The Jewish physician Abraham of Aragon's services were particularly sought for eye healing.

Late Middle Ages

Eye diseases and epidemics are documented in Jewish texts. Rabbi Meshullam ben Menachem of Volterra journeyed to Palestine in 1481, and his diary describes that in Alexandria Egypt in the summer months the air is so bad that people were blinded for periods of 6 or more months.

In 1488 Rav Ovadia Bartenura, the famous Mishnah commentator and bible exegete from Italy, writes in a letter to his brother that most of the inhabitants of Alexandria suffer from blindness.

Elia of Pesaro Italy in Cyprus in 1563 writes that Opthalmia is common in Cyprus in midsummer.

John of Aragon (1458-1479) employed a Jewish oculist named Cresques, who in 1468 operated on both eyes for cataracts. His operational technique of couching is found in Graefe's Archiv. The king was relieved of his blindness and lived another 11 years.

Vesalius described the anatomy of the eye in the 16th century and Kepler (1571-1630) described the physiology of vision in the beginning of the 17th century. The physician Amatus (1511-1568) wrote a treatise on ocular diseases, as did Zacutus Lusitanus (1575-1642) who described Gallic opthalmia. As late as 1816 Arthur Schopenhauer wrote a treatise on vision and colors based on discussions with Goethe (*Theory of Colors*, 1816).

Montalto a physician to Queen Marri de Medicis wrote a treatise *Optica Intra Philsophiae & Medicineae aream de Vistu de Visus organo & Objecto theoriam coplectens*, which was published in Florence in 1606 and in Geneva in 1613.

One important advance was Daviel's Extraction of Cataract in the middle of the 18th century.

Modern Enlightenment

Besides assimilation the enlightenment opened the doors to greater opportunities for Jewish physicians in medical education.

Marcus Eliezer Bloch (1723-1799) practiced in Berlin. In his *Medizin Bemerkungen* 1774 he described *colomba* of the iris with illustrations.

The Jewish physician Abraham Meyer in Hamburg published *Abhandlung und Beobachtung ueber einege Krankheiten der Augen* in 1785 (Fuerst, Bibliotheca Judaica 1851, vol 1, pt ii, 369)

George Hartog Gerson, the son of a physician, in 1809 wrote his thesis: *De forma corneae oculi humani deque singulari visus phaenomen.* It describes the dissimilar curvatures of the cornea in the vertical and horizontal meridians (astigmatism). His son Caesor became an opthamalogist in Hamburg.

Nathan Gerson Melchior (1811-1872) was an outstanding ophtamalogist in Denmark.

Karl Friedrich Canstatt (1807-1846) practiced in Ratisbon and published on ocular pathology

Samuel Mortiz Pappenheim (1811-1882) of Berlin published *Die specielle Gewebslehre der Auges* in 1842

Mortiz Schiff (1823-96) did numerous studies in opthamology including studies on the effect of section of the trigeminus on the nutrition of the eye, action, of atropine, and physostigmine and of the strychnine of the pupil, the pupil as aesthesciometer.

In December 29,1851 Dr. Remak and Dr. Abarbanel made the first reports on the use of the opthalmoscope.

Richard Liebreich (1830-1917) of Konigsberg made a simple opthamlmoscope. His *Atlas of Opthalmoscopy* is a classic.

Rudolf Berlin (1833-1897) was a pupil of von Graefe and prolific contributor to ophthalmic literature.

Moritz Schneller (1834-1906) founded an ophthalmic hospital in Danzig

Hermann Cohen (1838-1906) of Breslau published in ocular hygiene.

Julius Samelsohn (1841-1899) founded a private ophthalmic hospital in Cologne.

Jacob Stilling (1842-1915) was a professor of ophthalmology in Strassburg from 1884-to 1915.

Theodore Axenfeld born in Smyrna (1867-1930) became president of the *German Opthalmological Society.*

Friedrich Gross (1797-1858) founded an eye institute in Frosswardein and was chief physician of the Jewish hospital in that town.

Other important Jews who contributed to the field of opthamology up until World War II include: Ignatz Hirschler (1823-1891), Ludwig Mauthner (1840-1904), Isidor Schabel (1842-1908), Solmon Klein (1845-?), Wilhelm Goldzieher (1849-1916), Leopold Konigstein (1850-1924), Stephen Bernheimer (1861-1918).

In Russia Jewish opthamologists include: Emanuel Mandelstamm of Kiev (1839-1902), Leonhard Hirschmann (1839-1921), Miron Eliasberg (1865-?).

In France before the war opthamalogists include: Julius Sichel (1802-1868), Louis Laqueur (1839-1909), Henri Frenkel (1864-1934), Louis Emil Javal (1839-1907), Emil Berger (1855-1926), Zachariah Lawrence (1830-74)

From the above *desiderata* we can see that Jews in opthamalogy made an important and positive impact on this subfield of medicine. From Mar Samuel in the Talmud to the Rashba (Rabbi Solomon ibn Adret) who lifted the ban on studying science and philosophy caused by the Maimonidean Controversy, to Rabbi Simon Duran (1361-1444) who encouraged Jews to gravitate to medicine because of the opportunity to do good deeds and serve as instruments of Providence to bring *refuah* and relief from ailments, to the delegate to Napoleon's Sanhedrin in 1806 named Rabbi Dr. Graziado Neppi, Jewish physicians have always maintained a high priority to maintaining the health of the eyes, as vision is a most prized sense, although great Mikubalim of the middle ages such as Rabbi Yitchak Sagi Nehor (רבי יצחק סגי נהור)(Rabbi Isaac the Blind), is said to have seen with the inner eye that radiated prophetic vision. Indeed to our father Isaac son of Avraham, we see how poor eyesight can even lead to the transference of blessings, to Yakov and Esauv. Isaac in the bible was said to become blinded when the angels cried that he would be offered in the *Akedah* on *Har Mariah*. Yet a heavenly voice of the angel Michael cried out, "אל תשלח ידך אל הנער ואל תעש לו מאומה". Rashi importantly notes the text does not say "sacrifice Yitchak" but rather "bring him up" implying to a higher degree of holiness itself.[28]

---

[28] http://libguides.tourolib.org/ld.php?content_id=3994117

/

Chapter Four

The Doctor and the Newspaper Editor/Critic

Correspondence between Logan Clendening M.D.
and H.L. Mencken 1924 through 1944

What would a doctor and a news paper editor have in
common? Well Logan Clendening, a medical doctor from
the Midwest, Kansas City to be specific, had written an
engaging and somewhat provocative medical text for the
every day reader in 1924, Methods of Treatment, that H.L.
Mencken then the editor of the American Mercury found
well written, clear, crisp and much to his liking. Mencken
was born in Baltimore, an irreverent writer and critic,
influencing many of the writers of the Jazz age. This was
not the first time H.L. Mencken would go out on a limb to
encourage, support, and publish up and coming authors.
On Mencken's list of those whom he encouraged already
were Dreiser, Hemingway, F. Scott Fitzgerald, Willa Cather
and a host of others. Their writings would help fill up his
American Mercury and he enjoyed sharing the lime light
of their success. But a doctor book, whose syntax was
usually remote, stodgy, arcane and indigestible was
something else. Mencken and his publisher, Knopf were

looking for something new, such as a down to earth doctor speaking directly to the people and they considered that Clendening might just fill the bill, judging from the style and content in Methods of Treatment. This would be something new for the public. Medical reporting in a newspaper or magazine or an authoritative readable book directed to the people was unheard of at that time and just might make a hit.[1]

Just what did Mencken and Knopf, Mencken's publisher, find in Methods of Treatment to wet their appetites? First of all the book was laid out in clear divisions or headings; the first part included therapeutic procedures under heading such as drugs, diet, hydrotherapy and the second part considered their application and results to be expected in a variety of medical conditions. As Clendening would write in his preface, "Nowhere else, as far as I know, is such a plan carried out. It is the only justification I have for the book's existence." And then there were one liners such as in Chapter I under General Therapeutics / The Methods used in Treatment "Rest and surgery are the most effective therapeutic methods the modern physician has to offer the patient." Right off, that is an off key, bold, provocative statement. What about exercise and getting into shape? Or take this provocative two liner, "Surgery does the ideal thing – it separates the patient from his disease. It puts the patient back to bed and the disease in a bottle." Well this would not be your every day doctor lecturing to his patients and might tend to appeal to the unorthodox sharp witted Mencken. In *"Modern Methods"* Clendening

---

[1] A much larger collection of the correspondence between Theodore Dreiser and H. L. Mencken, 1200 litters has been published , Dreiser – Mencken Letters: the correspondence of Theodore Dreiser and H.L. Mencken, 1904 – 1945, edited by Thomas Riggins covering many of the same areas as the Clendening/ Mencken letters. Published by the University of Pennsylvania Press. 1986.

continues, "Immature faddists are continuously proclaiming the value of exercise: four people out of five are more in need of rest than exercise."[2] To top off this way of thinking…"Rest in bed will do more for more diseases than any other single procedure"

I am sure Mencken rollicked in this clear minded, mildly irreverent type of writing. In fact Clendening took on more of a Menckenist twist as the years went on. Many of the authors Mencken promoted would early on find a voice that mirrored their advocate. Finely to add some authority to his introduction Clendening quotes a beautifully freely drawn essay on the use of drugs written by William Osler.[3]

With Mencken's blessing and Knopf's publication this is how Clendening's *The Human Body* came to be published in 1927 to sell over half a million copies and set the stage for Clendening's newspaper column, *Diet and Health* to be syndicated in 383 news papers around the country, not to mention a dozen or more popular books on medicine and health.

Mencken's Role in Genesis of "The Human Body"

And so Mencken many years later, May 12, 1947 in a letter to Dorothy Clendening several years after her husband's death, responding to an article by Henry J. Haskell in the Kansas City Star at the eulogy of Clendening, would write

---

[2] Mencken would use three of these catching aphorisms in his Dictionary of Quotations. See correspondence about permission of Clendening to use these in his Dictionary on page 14.

[3] Osler, William, Reserves of Life, St. Mary's Hospital Gaz 13, 95-98, Nov. 6, 1907.

explaining his endorsement of her husband's writing efforts.

I fear that Haskell[4] in his speech rather over exaggerated what little hand I had in "The Human Body." What actually happened was this: I had an old and good friend here, Herman Shapiro, a very learned Russian Jew who served various Johns Hopkins men as translator. One night while we were all sitting at the beer table he showed me a book review in the Journal of the American Medical Association. It was a notice of Logan's Modern Treatment". The notice itself was commonplace but in it were several quotations from the book. Shapiro called my attention to the fact that they were beautifully written—a very rare thing in medical writing. I agreed thoroughly and sent for the book at once and when I read it I was delighted. Some time before this I had suggested to Knopf that a book on the order of" The Human Body" was sorely needed and would probably make a success. The next time I was in New York I told him about "Modern Methods of Treatment" and he also read it. The result was the proposal to Logan to do "The Human Body". It turned out to be a curious coincidence that he had had the same plan as myself. for some years past. He produced the manuscript quickly and Knopf and I agreed instantly that is was precisely what was needed. The rest everyone knows.

---

[4] Henry J. Haskell gave the eulogy to Logan Clendening on April 15, 1947 at the University of Kansas Hospital in Kansas City saying "Henry Mencken was the first to recognize the possibilities in Dr. Clendening's medical writings and persuaded Alfred Knopf the publisher, to invite the physician to write "The Human Body".

Preparation for Writing of *"The Human Body"*

Clendening's letter of January 2, 1926 concerns the congruence of ideas between himself and Mencken on the project of *The Human Body*. Note Clendening's conscious or unconscious use of Menckenese expressions as underlined, which become more apparent in his later letters:

Renouncing Alcohol, Tobacco, and World of the Flesh

Your letter outlines ideas which agree with mine concerning a book on the human body, as you doubtless have learned from a letter which I sent to Mr. Knopf and which he forwarded to you. I wish to express my appreciation of your interest in so far as it concerns me. Your ideas are very fructifying and I feel certain that I have conceived and intend to renounce alcohol, tobacco, the world of the flesh, and the devil during the period of my pregnancy until I shall be delivered.

I enclose a copy of a letter which I have sent to Mr. Knopf and which, in you capacity of male parent of this book, I feel you should look over.

Clendening sends this next letter to Knopf, having referred to a copy of it in the last letter to Mencken, indicating his agreement to start working on the book but has "several technical matters (which) must be settled first."

Mr. A.A. Knopf

750 Fifth Ave.

New York, N.Y.

Dear Mr. Knopf:

I shall consider it settled that I begin on the book on physiology right away. Mr. Mencken's letter has explained things very satisfactorily.

Several technical matters must be settled first.

1. How big should the book be? I do not believe it can not be made smaller than two hundred thousand words, and perhaps between that and two hundred and fifty thousand. Does that meet your idea?

2. Title. This may be premature but I am like a patient I had long ago when I did obstetrics   she said, if I just knew what to name the baby I think I could go trough this pregnancy more comfortably" "*The Human Body*" appeals to me but it was appropriated long ago by Martin, formerly

professor of physiology at Johns Hopkins.[5] That book, however is out of print. I do not know what the copyright laws are on the matter.

3 .The function of reproduction is one of the main functions of the body. How do you want to handle the subject? I spoke to a retail bookseller the other day about this book, asking him if there was much demand for such a work and what he thought should be included. His answer was, I think if you would write one on sex or sex life it would have a big sale. I think I can make up a chapter if you want it included.

4. Illustrations. I think it should be rather fully illustrated. They should be simple pen and ink drawings and will do more, if properly planned, to sell the book to the person who picks it up from a book stall than anything else. Do you want me to employ an artist here or send on sketches for one of your artists? How shall the expense of illustration be apportioned and what price per each do you consider reasonable if I employ the artist?

---

[5] H. Newell Martin was the first Professor of Biology at the newly established Johns Hopkins University. He was from Ireland and trained under Thomas Huxley, who recommended him for the position. He was interested in the newly developing field of physiology and made significant contributions to this field performing a variety of studies with vivisection and organized the department with a number of post graduate students who were to become distinguished. . His book *The Human Body* published in 1890 was unlike Clendening's which was primarily addressed to the lay public but rather was an elementary textbook on physiology for beginning students in the field, written a dry pedantic scholarly fashion, but going through multiple editions liked Clendening's book of the same name. Interestingly Martin's career was cut short by alcoholism and depression , much like Clendening's career may have been but complicated by a severe protracted peripheral neuritis. See *H. Newell Martin: A Remarkable Career destroyed by Neurasthenia and Alcoholism,* W. Bruce Fye, J. of the History of Medicine and Allied Sciences, Vol. 40, ps. 133 – 167, 1985.

These were all appropriate questions to be raised by the author to the publisher. Note the fumbling around with the problem of sex and how it should be presented, mirroring the mores of the 1920s. This will be discussed later and I hope you won't be disappointed in Clendening's handling of sex. Illustrations on the other hand, as we shall see, proved to be a strong feature of the book but note concern with expenses throughout.

A general outline of the book accompanies the next letter of January 21, 1926, nineteen days after the above letter to Mencken. Clendening writes to Mencken reviewing again the plan for the book. He indicates the book is "marching along pretty well. I hope to finish the damn thing by April 1st. It gives me no peace. Day or night".

The Book is Marching Along Pretty Well

January 21, 1926

Dear Mr. Mencken:

How is the enclosed for a plan of the book? You see I am trying to work out a presentation of disease which will get the layman away from the idea which is so firmly imbedded in him, that disease is a strange new lawless something which enters into his body from the outside. Your notice, I have a preliminary chapter on the healing of wounds, etc. The layman thinks that a sore hand, i.e., a healing would is a disease process. He thinks that enlargement of the heart is a disease process. Of course it is a compensatory reaction. In the last chapter, I take up the actually degenerative diseases, such as neoplasms. What do you think of this idea?

The book is marching along pretty well. I hope to finish the damn thing by April 1st. It gives me no peace. Day or night.

Has it ever occurred to you that it would be a good thing to advocate the passage of an amendment making it obligatory upon every citizen of the United States to take three drinks every day of the calendar year?........ Just as many medical and social arguments can be marshaled on one side as on the other. I t strikes me that the idea rather fits in with some of your recent encyclicals

Very truly yours,   Logan Clendening.

Clendening and Mencken seem to share similar view on the drinking of alcohol, making an amendment, presumably to the Constitution, obligatory for every citizen to take three drinks every day. This will be a matter of discussion in future letters.

Next is a letter from the publisher, Alfred A. Knopf to Mencken, dated March 15, 1927 indicates considerable uncertainty on the part of Clendening as to the title of the book:

Another Title for the Book? Tom Beer / Marie Stopes

Dear Henry,

Clendenning (Clendening) doesn't like "the Human Body" for a title any longer and would like to call his book The Paragon of Animals, which amuses me so much that I am almost afraid of it.

He argues that Tom Beer's book[6] wouldn't probably have sold nearly so sell as it did if it were called The Nineties or something of that sort and that there are so many dry books like Marie Stopes' called The Human Body,[7] that he ought to avoid that title if he possibly can.

Will you let me know your reaction by return as I won't be able to reach him after Saturday. He leaves for Europe Saturday morning. Yours, (AA.K.)

The letter of Clendening quoted earlier indicated some concern over the title but it is presumed that Mencken mediated the appropriateness of the title.

The next letter of September 27, 1926 to Mencken indicating considerable progress on the book. Clendening seems to have scaled down the "two hundred thousand words" he originally proposed to "about one hundred thousand words, an improvement over the two hundred and fifty thousand which so properly appalled (spelt as

---

[6] Thomas Beers published "The Mauve Decade" in 1926, a colorful rambling book about the 1890s The 1990s were sometimes referred to as the "Mauve Decade" because William Henry Perkin's aniline dye allowed the widespread use of that color in fashion. He took the title from a quote from artist James Whistler: "Mauve is just pink trying to be purple."

[7] Marie Carmichael Stopes, a Scottish lady interested in woman's rights, birth control, role of women in marriage, etc whose book by the same title as Clendening's, "The Human Body", treats the subject from a more sociological point of view. Compare her opening chapter "The Individual Units" – "We human beings always thing of ourselves a whole individuals...A dismembered human being is no longer able to play his part in this world." Compare this with Clendening's opening: "The human body is an animal organism, differing in only a few respects from other animal organism, and fitted, by the processes of selection and evolution for the performance of two main function: 1. Conversion of food and air into energy and into tissue, 2. Reproduction of the other individuals of its species."

appaled) you and Mr. Knopf." In addition he writes of the problems with the illustrations in the last paragraph:

Sending You Portions of the Manuscript: The Illustrations

Dear Mr. Mencken:

Under separate cover by express I am sending you portions of a manuscript "The Human Body" about which you and Mr. Knopf and I talked when I was in New Your last spring. I send you these for your suggestions and advice.

The entire manuscript is completed and about three quarters of it revised. Most of it is being retyped after my revision. I have got it into about one hundred thousand words, an improvement over the two hundred thousand which so properly appaled [appalled] you and Mr. Knopf. [In an earlier letter Knopf and Mencken had balked about the suggested two hundred and fifty thousand words.]

I am also sending Mr. Knopf some of the illustrations for his consideration and yours. I am having more delay is getting the illustrations completed than anything else. There will be about eighty of these in all.

Thank you for all your trouble and interest. I remain, As ever, Logan Clendening

Clendening was right, the illustrations are the heart of the book and along with the terse writing make it the best seller that it was to become. The eighty or so illustration in the initial edition and not one to many. The best are the pen and ink drawing, just as Clendening had indicated.

Consider Figure 2, Vesalius obtaining his first skeleton outside the walls of Louvain – a spooky rendition of a skeleton hung from a scaffold with a raven perched on its right shoulder, remains of a fire beneath and Vesalius, robed, sneaking up, but looking behind, to snatch the body.   Or consider Figure 57 "the Lovesick Maiden" depicting the handsome young physician examining for purposes of diagnosis the maiden's urine contained in a flask. She appears more interested in him than in the flask or the outcome of the verdict on the urine. Clendening points out that such "Liebeskranke Madchen" are common in the galleries of Europe, this on from Buckingham Palace, and he was impressed by the wise look of the doctor.[8]

The Presentation Copy: Comparison of Mencken to Horace, Isaac Walton, Mr. Pepys, Lamb and Thackeray

The next letter, combining the original handwritten letter and a copy, undated other than 1926 to Mencken indicates completion of the project and congratulates Mencken on the "Fifth Prejudices" (one of Mencken's many publications of essays from his journal articles). Note the informality of the greeting, "My dear fellow"

1926

The Presentation copies arrived and it was gorgeous of you to send them. I shall save them until that inevitable time arrives when I am old and impoverished and shall sell them for enough to keep me in luxury and catheters during my declining years. [unfortunately Clendening never got to enjoy those declining years]

---

[8] Clendening indicates that this "Liebeskrunke Madchen" is by Gerard Dou at Buckingham Palace.

The Fifth Prejudices is the best of the lot- you grow mellower. The great glorious thing about writing is that it reveals the personal-That's all that ever made any writing live- You will remain with Horace, and Isaak Walton and Mr. Pepys and Lamb and Thackeray and all the others on account of their intimate realizations.

When next you arrive we shall hale you not as a celebrity but as a friend.

Clendening places Mencken in a star studded cast of authors, perhaps to get into his good graces or more likely he really enjoyed Mencken's writing. His knowledge of Mencken is also apparent and his use of language increasingly mirrors the Menckenese. Unfortunately because of Clendening's untimely death he was not able to use the Presentation copy to "keep him in luxury and catheters during his declining years."

The Pratt Library in Baltimore has a copy of the first edition of "The Human Body" and contains a hand written dedication to Mencken, "To Henry L. Mencken with a favorable prognosis from (signed) Logan Clendening. I am not certain whether the dedication refers to the health of Mencken or the book itself.

Assessment of "The Human Body"

So how does "The Human Body" come off as a reading today? There was no Dr. Phil on television , no one minute medical report on the radio, and no syndicated columns on diet, back aches and sex problems. Only patent medicine advertisements to mislead the public. It was into this void that Clendening, Mencken and Knopf hoped the book would occupy in the mid twenties.

What are the distinguishing features of The Human Body as examined today? It is written in an easy to read, direct, often humorous vein, light and airy, bouncy style. The language keeps your interest, often punctuated with catchy phrases and twists and turns. As an example consider his characterization of Gallen:

> Galen, who with Aristotle shared the throne of scholastic authority in the Middle Ages, belonged to a late century and a different era. It was the brazen commercial world of Graeco-Rome. Galen was born in Pergamos, he lived in Rome, and he wrote in Greek. He was perfectly typical of his age. Indeed, his age being much like our own, he was not unlike a fashionable physician of today. He was shrewd, meretricious, garrulous, boastful, incredibly disputatious, and more than occasionally correct. His writings are so voluminous as to make the Holy Bible look like a pamphlet.(Mencken would have given his thumbs up at that.) He treated emperors, courtesans, wine-merchants, generals, senators, vestal virgins, oriental-rug dealers, philosophers, and gladiators. He tells of all his patients, and of the little tricks he used to arrive at his diagnoses. His case histories undeniable have a flavour.
>
> He specialized in omniscience. He was a second-rate genius, exalted by the pedants of Oxford and Padua to the status of a god. His knowledge of anatomy was learned entirely from dissections on animals. And for twelve hundred years the brilliance of his rhetoric prevented men from learning at first hand anything about the human body.

Or consider Clendening's account of the William Withering's introducing digitalis into medicine. Clendening relates that Withering learned of an old woman in Shropshire who brewed a tea of herbs which was used for dropsy. Clendening continues:

> Withering's next step was to obtain the recipe of the tea. I imagine that this was not so easy. The old woman unquestionably made a good living from it. She was attended by an enormous reputation. She probable guarded her secret very, very zealously. I like to dwell upon the possibly sinister devices with which those two compact and stubborn intellects conducted their contest- he to get, and she to keep. Bribery is more than likely; theft far from improbable. It is insipid to think that the business was achieved by the easy bamboozling of a simple old woman by a great gentleman alighting from his coach, clinking a pair of gold pieces in his pocket.

Now that is the kind of glib supposition not frequently found in a text book of medicine and sounds more like the spicy prose of Mencken.

A Contemporary Book Review –At last, A "Family Doctor Book " that is Honest by Raymond Pearl[9]

The first several paragraphs of Dr. Pearl's review of Clendening's "The Human Body" explains the appetite in

---

[9] Raymond Pearl was a distinguished Professor of Biometry and Vital Statistics and head, Dept of Statistics at the Johns Hopkins School of Hygiene and Public Health 1923- 1940. He wrote extensively on a wide variety of subjects and was the first to note association of smoking cigarettes and death, decades before the Surgeon's General report on association of smoking and cancer of the lung. He also had evidence that alcohol did not shorten life. He was a member of Mencken's Saturday night club in Baltimore, a group of prominent citizens and physicians who played music until midnight and then drank alcohol and ate to their heart's content. The influence of Mencken on Pearl's favorable review is unknown.

American homes for a doctor book, he explains that in the rural setting of most early Americans, without easy access to physicians there is a need to have such information on their shelves, " even thought it is never opened, similar to that of an African amulet protecting them from the necessity of worrying about various things". A second reason that made such a book a "staple of the book agent's wares during all the years was that the Puritan was never able to rid himself of the notion that sex was sinful, requiring a "family doctor book which was God's schedule of rates and tariffs for this department of spiritual trade." Pearl now comes to a consideration of Clendening's book itself, announcing "the glad tidings that at last there has been written a sound, readable witty, humorous, beautifully printed and illustrated and best of all, an honest book about the human body in health and disease. The performer of this latter day miracle, Dr. Logan Clendening, of Kansas City, is officially a practicing physician, but he is also a human being who has wide knowledge and deep understanding of many things besides medicine.---the manifold and comprehensive material which the author manages to present in a single volume is organized into four main divisions: 1. the human body as a unit, 2. as an organism for the conversion of food and air into energy and into tissues, 3. an organism for the reproduction of its own kind and 4. the human body and disease.---This third part describes the anatomy and physiology of human production and gives the sanest advice to the young and old on the problems of sex that I know of anywhere in print.. The fourth part discusses some of the more important diseases to which mankind is liable, and shatters a multitude of superstitions some of which are very orthodox and are to be met in the highest circles of medical society. The reviewer is in agreement with Clendening that the routine yearly physical examination is nonsense and may lead to mischief. The reviewer quotes the author who describes such a patient who has had a yearly examination who "begins on a diet to reduce his uric acid, is denied whisky , gin, tobacco and venery. In fact nothing which might lighten the gloom. Not one of his abstentions changes the tissues of his body. If he had not had the examination he might have lived twenty-five years without a symptom. He has been turned from a happy, self contented member of society into a morose, apprehensive hypochondriac. Etc ,etc.   Dr. Pearl finds some criticism in the author's handling of heredity, the reviewer's expertise. Over all the reviewer concludes, "By and large the book is a notable achievement from the literary, the scientific and above all, the human point of view".

Mencken, not to be outdone, also reviewed the book in The American Mercury, as the designated book reviewer for that magazine:

Mencken's Review of *The Human Body*

in *The American Mercury*

There is in this somewhat formidable tome a mellow and amiable wisdom which lifts it far above the lever of its kind. Most medical men write atrociously, especially when they address the layman, but not Clendening. It is by long odds the best work of its kind that has yet come to light in America.   It is devoid alike of the evasions and obfuscations which make nonsense of the school physiology books, and of the propagandists fervor which botches most of the handbooks for older and more wicked readers. What he has to say about sex, for example will probably cause uproars among the sex hygienist.......And what does Clendening have to say about sex?

Under "The Male and Female Organs of Reproduction," Chapter ll Clendening states:

> More and more frequently as time leaves the tracery of respectability on my features, as the excess of my youth tend to be forgotten and my figure grows magisterial, my friends are accustomed to send me their children at about the age of adolescence in order that I may explain to them what is called the secrets of life. I sit and look at these preternaturally solemn young persons, primed by their parents or guardians to expect some esoteric and recondite lore, and I wonder what I am going to say......... And so, quite frankly, I simply dodge the main issue.....When I attempt to elaborate some of my reticences, the parents protest that unless the children learn it from me. They will learn it in the gutter. My reply to that is: "The gutter is a very good school" I am myself a matriculate from the gutter.

Thus Clendening's "The Human Body" was sent off with a favorable prognosis from the reviewer after much attention by its god parents, Mencken and Knopf. Raymond Pearl's review was only one of at least forty reviews of "The Human Body" ranging from the Yale Review to the DeMoines Register, including one just reviewed from Mencken himself.[10]

---

[10] American Mercury, 1927, "*A good deal of the charming humanism of the late Sir William Osler is in him He writes gracefully and clearly, and he never forgets that the human machine is also a man, ant that the man has hopes and dreams as well a liver and guts. Most medical men write atrociously, especially when they address the layman, ,but not Dr. Clendening. It is by long odds, the best work of its kind that has yet come to light in America.*"

Request from Mencken for Contribution to the American Mercury

Even before "The Human Body" was completed Clendening was to receive from Mencken a request for a contribution the "The American Mercury". Clendening's reply in an undated letter from 1925 follows:

Dear Mencken :-

I am very much flattered that you have written asking me for a contribution to the American Mercury. It did not need your description of its habits and nature to make me acquainted with it as I have in my intimate book case every issue since Volume 1. No. 1.

I will turn over the matter of you suggestion in my mind for a few days and send you a manuscript at my earliest convenience. I have no particular inspiration about an article on drugs but I have something else in relation to medicine which has been strongly on my mind for some time and which may suit your needs.

Sincerely yours,   Logan Clendening

Clendening shows his familiarity with Mencken's *oeuvre*, having every issue of The American Mercury from the very first issue. Clendening's article prompted by the above request eventually found its way into the journal in Volume XVll of May 1929, p. 54, under the title Meat. This is a seven page diatribe on the virtues of protein in the form of meat against the proponents of the vegetable, fruit, and nut diet. (He denounces the effect of meat elevating serum uric acid and being the cause of such conditions as hypertension and Bright's disease. Clendening concludes, "man instinctively is a meat eating animal. His ideal diet, of course, is not one sided in any direction. It is a mixture of all kinds of foods- meats, fruits, vegetables, fats sweets, nuts, salts, bread and water- fresh and preserved, cooked and raw.")

Clendening would go on to write a bundle of popular and medical books such as Methods of Diagnosis, Source Book of Medical History, Behind the Doctor, Methods of Treatment, The Balanced Diet, The Care and Feeding of Adults with Doubts about Children, as well as a charming A Handbook to Pickwick Papers.[11] In addition to "Meat" Clendening would contribute a number of articles to the American Mercury between March 1926 and May 1929 including one on "Drugs" None of these endeavors however had the prodding or editorial review of Mencken or Knopf as did the genesis of "The Human Body", at least as judged by letters between them. Knopf however published a number of these books. In addition Clendening went on to write a syndicated medical newspaper column, Diet and Health, for many years. Clendening became interested in collecting books on the history of medicine on many of his trips abroad and

---

[11] The preface to this book describes his introduction to Dickens: *It was my mother who decided I had had enough of Horatio Alger, Jr .Louisa M. Alcott, etc, and should begin on the real "classics". She was wise in her first selection, for her choice was "Nicholas Nickleby". I was sent to the public library to get it...It was a squat one storey brick building on Eight Street between Oak and Locus in Kansas City. I took out a card and wrote my request and soon a very dirty copy of Nicholas Nickleby was shoved through the window to me. I was suspicious of it at first. I was suspicious of it at first, but on the way home I glanced through the pages and was apprenticed to the "classics" for ever.* Mencken had a similar experience when he visited the Enoch Pratt Library branch in West Baltimore. He found a copy of Mark Twain at home on a top shelf and on reading found that "I had entered a domain of new and gorgeous wonders and pressed on to the last word." ("Mencken-The American Iconoclast" Marion E. Rogers, 2005)

donated them to the History of Medicine Library eventually named after him.

With all this activity he reduced and eventually discontinued his medical practice giving his time to writing and collecting books. However the letter writing exchange between Mencken and Clendening went along for more than twenty years until the latter's death in 1946. Their topics of conversation while occasionally related to medical subjects include the state of their health, getting older, Mencken's hay fever as well as congratulation to Mencken on the his approaching marriage in 1930. There is a beautiful letter from Clendening on the death of Mencken's wife, Sarah, after only four and one half years of marriage. This letter of condolence is to b e matched by the stark compact letter to be reproduced later on page 20 by Mencken to Clendening's widow in 1944 at the time of the death of Clendening. 12 There are also requests of Mencken to Clendening on the meaning of certain expressions to be added to his American Dictionary, such as what Clendening thought was the origin of the expressions Charlie Horse and glass arm. The subject of alcohol comes up a number of times.

Clendening's letter tend to be long winded, hand written or typed, Mencken's letter always typewritten, brief, succinct, crisp and with sardonic humor.13 Clendening's letters often have the tone of a physician advising Mencken to slow down, don't answer so many letters,

---

12 Mencken would have a stroke in 1948, leaving him unable to talk or write and died in 1958.

13 Menken was known for his personally answering all of the voluminous letter sent to him. His well worn miniature by current standards, Corona typewriter is on display in the Mencken Room at the Pratt Library in Baltimore.

describing his own gout, etc. Some examples of these exchanges follow:

Clendening's Advice to Mencken: "don't answer this letter no try to keep up with this gigantic correspondence we fools thrust upon you."

Clendening to Mencken 12/22/1927

Clendening seems depressed, expressing concerns about getting older Clendening was 43 years old at this time and four years younger than Mencken). He offers Mencken advice on conserving his energy, the advice sounding more like measures he would like to take himself. Clendening seems to have a good idea of the massive amount of Menken's correspondence with the phrase, "don't answer this letter nor try to keep up with this gigantic correspondence we fools thrust upon you."

> Dear Mencken: I am sorry you can't stay with us in June but was afraid that you would have to stay with your crowd. However we are delighted you will be here and hope that you will look us up as soon as you arrive.
>
> I was very much distressed to hear from Beck that you had decided to give up the column in the Tribune for a time and that you felt you were overworked. Take care of yourself. You and I are about the same age; if anyone talked to me ten years ago about using up reserve strength I would have laughed at them. I thought I could do anything, eat and drink anything and show no effects. But in the last few months I have found I was getting tired in the afternoon that I just can't do comfortably. By God I'M (I am) getting old.

That's the fact. Bank your fires a little. Go on the wagon for a month just to see if you don't feel better. Rest and walk aimlessly every day. Let the wells of life fill up in you Lie fallow a while. Don't answer this letter nor try to keep up with this gigantic correspondence we fools thrust upon you.

Though I must say you show no evidence of diminished powers. How anything could be better than your review of Judge Lindsay's book on marriage[14] I don't know. You know I had to be fitted for glasses the other day. God Dam it. And I turned down a perfectly good new invitation to the dance the other evening. It' s hell but we must face it with much fortitude as we can.

---

[14] Mencken reviewed "The Companionate Marriage" by Ben B. Lindsey in the "American Mercury, . The author argues to give women more rights such as use of contraceptive devices, granting divorce if mutually desirable and denying alimony to a woman without children and capable of earning a living. Mencken thinks these measures may be helpful but he insists that any scheme of marriage or any scheme of non-marriage is unlikely to avoid unhappiness. "The unmarried, on blue days, are miserable lonely and forlorn, and the married, on blue days, have too much society." He concludes pessimistically that "unhappiness in some form or other is the universal lot of man... The holy estate, even under the medieval rules that now prevail, has its moments of ecstasy, and, what is more important, its long stretches of solid contentment, but it also has its times of war, its times of intolerable irritation, its times of cruel dullness. Thus it is bound to be full of unhappiness, as any other estate of man is full of unhappiness." This is from a man who had derided the state of matrimony in the past but who would become very happily married in 1930, but its bliss would last only 4 1/2 years until the death of his wife secondary to tuberculosis meningitis.

Lustige Weinnachten[15] (Merry Christmas)

Logan Clendening

Request for Another Book

Next is a letter of August 17, 1928 from Clendening to Mencken which in part discusses Clendening's ruminations on a new book requested by Knopf in which his concerns on what to put in the book sheds some light on his thoughts for further literary endeavors. It also indicates the interest Knopf and Mencken have in encouraging Clendening to produce more popular medical topics to present before the public.

Dear Mencken:

I am going to impose on your good nature and ask for advice......I need help concerning a new book. Knopf said he would like to have me submit to him another manuscript and of course, I have already given the matter some thought. The trouble I have had is with a subject and theme. It would be easy to get up a sort of pot-boiling affair and call it *"Magic and Medicine"*, but I hate to get into a rut and look as if I were repeating. Just the other day I got an idea and it is this I lay before you.

---

[15] Clendening's use of the German "Lustige Weinnachetn" is somewhat irregular. The usual greeting for Merry Christmas is "Frohe Weinnacken". Lustige is just a little too strong, as in Lustige Witwe (Merry Widow) or in Till Eulenspiegel's Lustige Streich (Till Eulenspiegel's Merry Pranks). Either Clendening didn't know the correct German greeting or he was emphasizing the greeting to be more vigorous. My German consultants think it is the former, since how would a person from the Midwest really know how to use the German language.

As I conceive it is in the form of fiction and is to be entitled "And Then He Lost His Health". It purports to describe a gent who has a vague little neurosis going all over this magnificent land from famous clinic to famous clinic, and to dentists, allergists, endocrinologists, focal infectionists, gastro-enterologist, surgeons, and last, but not least, psycho-analysists (psychoanalysts) -trying to get a diagnosis and some treatment that will relieve him. He goes to the Mayo Clinic, which will be thinly disguised, and several places out herein this glorious land of California(God save the poor patient who gets sick in California), and so forth and so on. My head is teeming with incidents to illustrate everyone of the gentleman's adventures, I have thought of putting them into essay form, but they look too raw that way. The general subject-that of medical economics, especially in the United States – is of all subjects the most pressing one for medical writers to write about today. What do you think of the idea, and do you think my powers are adapted to it?

Yours very truly, Logan Clendening

In this letter we see Clendening's requesting advice from Mencken about a publication. He is anxious that he not repeat himself. The subject of medical economics seems as much a concern then as today. Mencken's reply is not available but we can imagine his response.

Congratulations on Mencken's Approaching Marriage

On August 7, 1930 we come to a beautiful letter from Clendening to Mencken who has just announced his engagement in marriage. Mencken was probably the most eligible bachelor of the twenties, proudly wearing his bachelorhood on his sleeve and writing about it. He did have some brief affairs with Hollywood types of girls that was rumored about, but his whole demeanor and writing never let on that he would be bitten by the matrimonial bug. But Sarah Haardt was one of his types, in the English department at a local girls college where he gave an informal talk each year and whose writings he had already published, encouraging her literary career. The letter to Mencken follows:

On Board S.S. "Homeric"   Thursday August 7

Dear Henry:

The Paris edition of the New Your Herald informed me of you approaching marriage. I hasten to offer my very heartiest congratulations. I think is a splendid thing for you and I am sure Dorothy as well as myself will be genuinely delighted.

I have had a very good experience in the holy estate of matrimony and can sincerely recommend it. I am sure you are excellently adapted to its rigours as well as its happy moments.

With the best possible wishes for your happiness and the happiness of your bride.   Logan

"Happy moments" seem to pale with its rigors. Whether he was "excellently adapted to its rigors" was uncertain considering his reputation as a bachelor and everything he wrote that went before it, but the marriage was a splendid affair but unfortunately lasted only four and one half years, because Sarah died with tuberculosis.

The following is a beautiful handwritten letter of condolence from Clendening to Mencken, undated but shortly after Sarah's death in 1934:

Clendening's Response to Death of Mencken's Wife

Dear Henry:

I was in Paris when the newspaper gave me the sad news of the death of you wife. I deferred writing you until I returned and then Alfred (Knopf) told me you were sailing on the Bremen on her return trip from our crossing. Which accounts for this delay in proffering my sympathy no less real and heart felt for that. I do not know what such a loss would be like but realize it is no place for the intrusion of anything but the proffering of my continued regard and affection and hope for your early acquisition of equanimity. Yours, Logan

The Mencken Clendening correspondence continues on a regular basis discussing a variety of topics from deciding on what topic to write about next, alcohol, health issues, questionable practices in the medical profession, reminiscences about boyhood experiences about a pony as well as the meaning of certain expressions for Mencken's forthcoming Supplement to the American dictionary, etc.

Clendening's Medical Problems and Advice on Topic for Next Contribution

The following letter from Clendening to Mencken dated December 1, 1931 relates to illness, a contribution to the American Mercury and indicates Clendening's shift from medical practice to writing:

Dear Henry:

My convalescence has been very slow and I suppose I have been very impatient about it but I can see that I am getting better. My trouble has not been mental or nervous reaction, which you implied in your last letter, but the fact that the damn place won't heal up. I have just within the last few days gotten a complete layer of thin skin over the wound and there are two or three places that either due to the fact that the scar is pulling or the nerves are spreading out, are as tender as a prima donna's feelings. Naturally, my thoughts have not dwelt very much on belles letters as I can barely sit up long enough to write what I have to.

You ask me about a contribution to the Mercury which, of course, I would love to do...... As I am not very closely in touch with any research work or clinical work any more I hesitate to suggest a scientific subject, and besides that there is so little that is very exciting in new medical discovery lately that it would be hardly worth while.

What would you think of an article on some reminiscences such as Chicago as a medical center from 1895 to 1910? I remember all the big men of that generation well and how we used to discuss Murphy's operations and Sippy's clinic. It would evoke the memories of a day that apparently is definitely gone in the teaching of medicine. It might be extended to the entire Middle West of that time, including the birth of the Mayo Clinic and some of the famous figures in Omaha St. Louis, etc.

The only other subject I have would be on social and economic medicine, something to the effect of "private practice vs public health practice" and would consider the encroachment on the legitimate field of private practice that the Utopian public health administrator of our day is making. I hope you and yours and all my friends in Baltimore are well.

The first paragraph of this letter seems to put Clendening in the role of a patient with Mencken as a doctor suggesting an alternative diagnosis to his condition. The social and economic aspects of medicine seem to be very much a part of their concerns. And that expression, "the scar is pulling or the nerves are spreading out (and) are as tender as a prima donna's feelings", is delightful.

Mencken's Review of Article on Leeuwenhoek and Clendening's Rejoinder

This next letter of Clendening to Mencken of January 4, 1933 on Leeuwenhoek, the inventor of the microscope, is a little lesson in optics and criticism directed towards the editor of the American Mercury:

Dear Henry:

In your review of Dobell's life of Leeuwenhoek in the January "Mercury"16 you betray some misunderstanding about his microscopes which you evidently derived from the author himself. You express surprise that he saw so much with such poor instruments, and suppose that in order to light his object he used a dark field illumination. Also that his secretiveness kept him from describing his methods.

---

16 Mencken regularly review books in the American Mercury.

The next time you go to Munich and visit the Deutsches Museum, one of the most magnificent sights in Europe, you should go to the section on optics and find the case illustrating the evolution of the microscope. There they have a Leeuwenhoek microscope mounted so the visitor may see the specimen. (I imagine that this is one or Leeuwenhoek's own original microscopes. Dr. Dobell mentions that there are two other genuine instruments now in Germany- one in a well known museum.")

The astonishing thing is how much you can see through this inadequate looking little instrument. There was what looked like some pollen grains or vegetable chaff mounted and I should judge that the magnification was at least a quarter as large as the low power of a modern compound microscope. As for the illumination it was brilliant from an ordinary electric light bulb. On a sunshiny day in Leeuwenhoek's Holland there would be equally brilliant light. He did not divulge his secret because it never occurred to him that there was any. My guess is that Leeuwenhoek depended partly for his magnification on having an extremely tiny hole for the eye piece.... Etc.etc.

Here the doctor is in his field and seems to set straight the editor. Mencken's prompt and gracious reply three days latter, January 7, 1933, follows. Considering the time needed for the letter to travel from Kansas City to Baltimore, this apparently is swifter than currently possible and only cost 2 cents :

Dear Logan:

I followed Dobell without challenging his facts, for I assumed that he had made an adequate description of the surviving Leeuwenhoek microscope. Some time ago I was talking about the matter to Dr. William H Welch (Dean, Professor of Pathology and History of Medicine at Johns Hopkins) and he expressed the opinion, as you do, that Leeuwenhoek could have seen all that he says he saw without the use of a dark field. The next time I may (be) in Munich I shall certainly take a look at the microscope there. It is, of course, not a microscope at all, but only a magnifying glass.

What are you up to, and when are you coming eastward again? The next time you see me I'll probably be selling apples in Fifth avenue.   Yours, H.L. Mencken

Mencken was known for his prompt reply to letters. He takes the criticism in stride. He also enjoyed the company of the doctors at the medical school like Dr. Welch and others, some of whom were in his Saturday night club. Selling of apples probably refers to the Depression years.

Clendening in Jail and Mencken's Note of Sympathy
February 11, 1939

This episode is very well known by probably every one. Clendening, upset by the noise from an air compressor drill near his home, takes an ax to it, spends four hours in jail, labeled by the newspaper as "Jailed after Nervous System Finally Rebels against the Noise.

In the Baltimore Sun of February 11, 1939[17] appeared an article with a bye line from Kansas City, "Author Takes Ax to Air Compression Drill On WPA Job and Stops the Damned Thing" The article went on to indicate that," An excited demonstration of the rebellion of man's nervous system against the tyranny of noise cost Dr. Logan Clendening, author of one-time best seller, "the Human Body" four hours behind jail bars today". The article went on to indicate that, "Dr. Clendening who conducts a newspaper health column, emerged from the ordinarily quiet study of his home in a fashionable residential section , descended on a WPA sewer project and wielded an ax with telling effect on the vital spot of an air-compression drill. His arrest on charges of disturbing the peace and destroying Government property followed. After four hours in jail he was release on $1,000 personal bond." The police patrolman reported that, " Dr. Clendening's complaint was somewhat as follows: 'Since last October 1, seven and on half hours a day, six days a week the air compression machine has given us its unwelcome symphony as the drill bit into rock and shale 100 yards from the physician's home' ". The physician was reported to have exclaimed," I am going to stop this damned thing for once and for all, waving his ax as he spoke." A picture of Dr. Clendening appeared in the newspaper of him behind bars with bowler hat, tie,

---

[17] Many additional accounts appear of this event including one by Jennifer Wilding in the Kansas City Star of February 9, 1986. Mencken also was also almost placed in jail over night in 1926 for deliberately selling a copy of the American Mercury banned in Boston allegedly containing a salacious article, "Hat Rack" on the Boston Commons protesting his belief in freedom of the press. Mencken was let off after the Judge having read the article ruled that there was nothing wrong with the article. He was supported by the Harvard students and Felix Frankfurter, then Law Professor. This event is best related in Prologue : Boston 1926, "Mencken, the American Iconoclast," Marion I. Rogers, 2005 Oxford University Press.

gloves, and carnation in lapel was labeled, "Jailed After Nervous System Finally Rebels Against Noise."

A short brisk rejoinder by Mencken, very typical of his style, follows on February 13, 1939, two days after the incident:

> Dear Logan:
>
> As a patriot and Christian, I offer you my felicitations. You have struck a blow for freedom that will ring down the ages. In case you are still in jail, let me know and I'll send you a Bible. I am only sorry that you didn't employ a pile drive instead of an axe and that you neglected to chain Harry Hopkins to the machine before tackling it.
>
> Sincerely yours, H. L. Mencken

Mencken apparently approved the act, "I offer you my felicitations. You have struck a blow for freedom, that will ring down the ages." Really ? And for the gift of that Bible while in the jail; actually he was fined $25.00 by Municipal Judge James H. Anderson on disturbance of the peace charges and $25.00 for destroying property and let go after four hours. Harry Hopkins was a member of Roosevelt" inner circle. Roosevelt was an anathema to Mencken and considered Harry Hopkins one of his henchman.

Request from Mencken to Clendening to Obtain Permission to Include Three   Quotations from "Modern Methods"

Mencken is planning to publish a Dictionary   of Quotations and in January 2, 1940 requests permission from Clendening to use several pithy expressions that we have seen previously that appeared in "Modern Methods of Treatment":

Dear Logan:

I have a letter from Mosby giving me permission to use those extracts from Modern Methods of Treatment that I mentioned in my letter of a week ago. They said however that this permission will not be good until it is countersigned by you. If it is in your heart to give it to me I'll be delighted. Moreover, I'll pray for you. Yours, H. L. Mencken

A straight forward request but prayer was not recognized as a strong point in Mencken repertoire.

A letter from Clendening three days later, January 5, 1940 follows:

Dear Henry:

Of course you may use any quotations you please from "MODERN METHODS OF TREATMENT (capitals are as per Clendening). I am flattered and proud that you wish it.

Your letter caught us only today as we returned from a holiday trip to Mexico. I share your feelings about Christmas; Mama (Dorothy, his wife) and I are both orphans and dread trying to make gay at holiday parties. In Mexico we saw some gawdy (gaudy) sights. Particularly the Virgin of Guadalupe – the only healing shrine I have ever visited. It would have been raucous had it not been so pathetic and disgusting. The way they bleed poor peons! Outside of the church they sell little silver votive offerings. There is a leg if your leg is sick, a hand if your hand is sick – an eye, an ear, a cat , a dog, a pig, a burro – and fin du siecle an automobile if you automoble(automobile) is sick . the Holy Church of Mexico is right up to date if there is any money in it. We also saw the Mayan ruins in Yucatan. Very impressive, and the accommodations for sight seers very comfortable. A civilization that was abandoned in the midst of the jungle and discovered only about thirty years ago.

I am consumed with curiosity about your anthology. I happen also to be doing one, "The doctor in Fiction, Peotry [Poetry]" ....{including a] fascinating chapter on doctors who have described their own diseases.[18]

You, sir have our Episcopal authority for the best of good luck in the coming year. Yours, LC.

Clendening was well traveled and this is his account of a trip to Mexico. He seems to be working on another book as well. He occasionally makes errors in typing or spelling, Mencken who never completed any formal training other than high school, makes hardly any errors at all.

As a follow up to his contribution to Mencken's "Dictionary of Quotations" Clendening seems to be proud of his entries but needing reassurance about "keeping the mind skipping around and alert", dashing off the following letter on April 14, 1942:

Dear Henry

I feel that I belong to the ages now that I am incorporated in "The Dictionary of Quotations". I

---

[18] Clendening was probably referring to "Behind the Doctor", published by Knopf in 1943.

had a lot of fun with you book last night – sat up in bed and read it until quite late.

I find at my age [58] that books like the Dictionary, "The Dictionary of Quotations ", "Who's Who in America", and a small encyclopedia are the best sort of continuous reading. They keep the mind skipping around and alert, which is very necessary as the arterial conduits begin to close down in the brain stem.

As ever yours,   Logan

Request from Mencken for Meaning of Three Expressions for The Supplement to The American Language

The American Language was Mencken's most scholarly work , a large tome on the history and definition of expressions unique to the American language, with two supplements.

A letter of November 11, 1943 from Mencken to Clendening finds the former trying to get some help on the meaning of three American expressions to include in his Supplement to the American Language, a compilation of American slang and expressions that were not included in the original edition to the American Language:

Dear Logan:

I have been trying to find out the true meaning of the American verb to goose - that is, I have been trying to find out what it is that makes a goosed man jump. Fishbein, of the Journal of the American Medical Association, has given me some help but refuses to print an inquiry in his Notes and Queries column, apparently on the ground that it would be an indecorum. To goose doesn't appear in any American dictionary but the National Safety Council and other such agencies have issued many printed warnings against the practice. It is not an uncommon occurrence for a goosed man to fall off a high scaffold and hit a couple of other men on his way down. Fishbein suggests that the verb may come from an old custom of testing a goose by feeling of the fat in the anal region. Have you any ideas on the subject? I want to print something about the word in a supplement to The American Language, under in progress.

Two other words that occupy me are Charlie Horse and glass arm formerly in wide use among baseball players. My impression is that Charlie horse indicated a serious rupture or strain, where as glass arm covered only a mild bursitis or synovitis. I can find nothing on the subject in the medical dictionaries, nor is there anything in the books at hand on tromatic (traumatic—one of the few misspellings by Mencken) surgery.

I surely hope that you and Mrs. Clendening are in the best of health and spirits. I am hard at work upon my book and have little time for anything else. Nervertherless, I manage now and then to get a really good dinner. I had one last night in the country not far from Baltimore. The excellent victuals were washed down with some prime Rhine wine. Yours, H.L. Mencken

Mencken was fond of seeking out the help of doctors for his inquires. We saw this with his approach to Dr. Welch on the matter of Leeuwenhoek's microscope. Ending with a mention of alcohol was not uncommon since both of them seemed to enjoy these two carbon fragments.

The reply from Clendening was prompt, voluminous and perhaps authoritative. Letter of November 20, 1943:

Dear Henry:

I am shocked and astonished that "to goose" finds no place in any American Dictionary. My only explanation as to why a goosed man jumps is that it is a reflex, but that is banal. As to the etymology – I have been trying to find proof of what my guess is. I seem to remember in some early woodcuts and as decorations on Greek pottery, a child with his rear exposed wandering around a farmyard, and a goose sticking its beak up into the gluteal folds. It's a habit of geese. The enclosed picture is a late Nineteenth Century corruption of these drawings, but the idea is there. (Don't return the page).

As to "Charley Horse". I can be somewhat more explicit. A Charley Horse is a ruptured muscle. All muscles are surrounded by a delicate membrane. Under stress this may tear and the raw muscle fibers herniated out through the opening. It is exactly the same pathology as "string halt" in a horse. There is your horse, but what genius added the Charley I have yet to learn.

"Glass arm" in pitchers is, as a term used by baseball trainers (medical), of more diverse nature. It may mean anything from a Charley Horse to a torn tendon, to a dislocation, bursitis, as you suggest and even small fractures.

I was once the close observer of a tragic little drama of glass arm. Etc, etc.

Miss Dorothy and I are living a monotonous, freezing life in our own home, getting nothing we like to eat, restricted in motor car movement – all to save the British Empire. I wish once the British Empire would conduct its own brawls personally itself.   Always affectionately, Logan C

Clendening seems very authoritative about baseball matters. His last gibe against the British in times of World War ll would appeal to Mencken's suspicions of the British.[19]

About Their Ponies': Frank and Dixie

In a lighter mood is the following correspondence between Dorothy, Clendening and Mencken on the matter of their experiences in youth with a pony in this undated letter, probably January 1943.

Dorothy having read "Memoirs of a Stable Boy", second chapter in the Day's Book, Heathen Days about Mencken's pony, is our next letter:

---

[19] Supplement One – The American Language by H. L Mencken published in 1945, the introduction states, "So much new material had accumulate after the publication of the Fourth Edition of *The American Language* in 1936 that a separate volume was deemed advisable to contain it. Though of particular usefulness in conjunction with *The American Language* this *Supplement* is nevertheless an independent work and may be read without reference to its predecessor. Pages 390 through 393 of this *Supplement* is a complete discussion including the sentence ,"Meanwhile, every American knows what *to goose* means, although the term appears to be unknown in England, and there are no analogues in the other European languages." Footnote 1 on page 391 lists Mencken's "indebtedness to the late Admiral Charles S. Butler, M.C. , U.S.N., to Logan Clendening, author of Modern Methods of Treatment, The Human Body, Source Book of Medical History, etc and to Dr. Morris Fishbein, editor of the AMA. I could not find a discussion of Charley Horse or glass arm in this *Supplement.*

Dear Henry:

Nothing in days, months, years, has delighted me as much as "Memoirs of a Stable Boy."[20]

We too, my sister and I, had a pony – by the name "Dixie". We had too, in those gay nineties days what went by the name of a "Governess Cart". Need I tell you – a square basketry job on two wheels with one rear door.

Well, the scene was laid in the wilds of Wisconsin and of an agreeable afternoon the sisters Hixon (her maiden name) would drive out complete with governess in said equipage – just about the hour when the Public Schools were "letting out". As we tooled along embalmed in our "Days of Innocence" cries went up from the home – going urchins-"there goes the Hixon was wash."

Well, one balmy summer day the circus came to town and Mademoiselle drove us to view the parade. All went well, and al were having a "dandy time" when along come the bugle call. At its first notes "Dixie" reared himself up on his hind feet and executed a "pas seule"[probably pas seul –a ballet expression indicating a dance for one]] the like o f which was entirely outside our bucolic

---

[20] Memoirs of the Stable appears in the third volume of Mencken's Days books, Heathen Days, Chapter 2, p. 15-26. Mencken wrote these books toward the end of his career, looking back on his childhood experiences. As Mencken says in the Preface to Heathen Days, "It is simply a series of random reminiscences, no always photographically precise of a life that, on the whole, has been very busy and excessively pleasant." The story is of a Shetland pony, Frank, who was boarded at the end of his long back yard in 1891 when Mencken was 11 years old. The pony played many tricks on Mencken and his brother, Charlie, the reading of which delighted Dorothy Clendening in the accompanying letter.

experience. We were tossed right out over the rear exit in a heap. Bystanders and circus attendants rushed to   our assistance- and the general consensus of opinion was that before "Dixey" was employed to haul the Hixon clothes basket he had served time with a circus.

God bless you for giving me a laugh in these grim days. Affectionately

"Miss Dorothy"   Alias Mrs Logan Clendening

Mencken responds in an equally light breezy rejoinder:

January 26, 1943

My dear Miss Dorothy:

I needn't tell you that I was delighted to hear about Dixie. I always suspected that my own pony, whose name was Frank, had some circus experience. He had some tricks that were certainly not learned in ordinary stables and he showed an actor-like delight in notice and applause. I hope that he and Dixie are now grazing upon artichokes and sugar in the heavenly pastures. In a few days you will receive from me a book in which "Memoirs of a Stable Boy" is a chapter. It is not to be published until March 1st, so say nothing about it.

I seize the opportunity to hope that you and the Mister are in good health and lively spirits. Please tell him that an occasional postcard from him would not do me any harm.

Sincerely yours, H. L. Mencken

The next week, February 2, 1943 Clendening writes a letter to Mencken thanking him for sending him the "latest volume of reminiscences" i.e: the Day's Books, "Happy Days, Newspaper Days and Heathen Days" as well as complaining about gout, difficult cold winter, increasing work in the medical school all perhaps harbingers of depression. His own experience with a pony is added:

Dear Henry:

It was thoughtful of you to send us your latest volume of reminiscences [the Day's book] which are, as always, amusing, particularly because so many of them coincide with my period of life and the pursuit of happiness in a middle-sized American town.

We are having a winter which – if I survive it – I will look back upon with horror. In the first place, I have had the gout twice. I don't think I have once been even reasonably warm in my own home and I am too old to have frets put upon me by those in authority. I have, naturally a great deal of increased work in the medical school which is not unpleasant. The only thing that sustains me is that I am writing a book of monumental proportions, which already dwarfs the *Encyclopedia Brittanica* [Britannica] and the collected sermons of Bossuet. It is on the subject or diagnosis and I am putting down everything that everybody ever said on it. It will probably never be completed – let alone published – but at least it keeps me busy from nine in the morning till noon.

You have given Miss Dorothy great delight by directing two letters to her, but particularly by your account of the Shetland pony. I had one too and the little bastard used to try to rub me off against telegraph poles and trees.

As ever yours,   Logan

Report of Clendening's Death and Mencken's Reply

The following report appeared in the Baltimore Sun, with a by-line from Kansas City, January 31, 1945, Dr. Clendening Found Dead, Authorities Say by Suicide. " the body of Dr. Logan Clendening, 59, nationally known physician, lecturer and writer on medical subjects, was found in bed at his home about noon today, the throat and left wrist pierced in such a manner that authorities reported there was no doubt of suicide... The instrument to inflict the wounds was a good combination cigar, clipper and penknife. ...the spectacular Kansas Citizen had been morose and depressed lately, probably over ill health.... Dr. Clendening was best known nationally for his syndicated column, 'Diet and Health', in which he expressed many unconventional ideas of his profession....Known as a man of whims and sardonic wit, Dr. Clendening paid little attention to conventional hours and it was not considered unusual that he did not arise a t his usual time today...."[21]

Mencken's letter to Mrs. Logan Clendening in characteristic brief, terse, and type written follows:

---

[21] Many articles have appeared on Clendening's motives for suicide among the best is a long letter by Dr. Don Carlos Peete to Dr. Ralph Major in the archives of the Department of the History of Medicine dated 30 October 1985 as well as articles by Dr. Major, Bulletin of the History of Medicine, Vol Xlll, No. 2, July, 1945, and by Robert P. Hudson, Journal of the Kansas Medical Society, Vol. LXlX,p154 – 156 and 160.

I needn't tell you that I am shocked and sorry beyond words.

H.L. Mencken

Dorothy's letter to Mencken on Her Current Situation

An undated letter from Dorothy was sent to Mencken:

Dear Henry: - Since Logan's death I have sold the Kansas City and Santa Barbara houses and bought the one at the address below. San Marino is practically a part of Pasadena and near my aunt and Cousins and think I am going to love living here. It was too hard going on where Logan and I had had much happiness and living with the ghosts.

Perhaps you will be coming out for the racing at Santa Anita and will look me up.

Sincerely   Dorothy (Mrs. Logan) Clendening.

Postlied

Appearing in the Kansas City Star Newspaper on April 20, 1947 under the by- line "Fountain a Fit Tribute, Dr. Logan Clendening is Honored in a Memorial, Henry J. Haskell[22] sees in it a Representation of the Energy and Spirit of the Physician-Author" a report is given of "an address in the tone and spirit of the man it eulogized by Henry J. Haskell, in dedicating the Dr. Logan Clendening fountain and

---

[22] Henry J. Haskell was editor of the Kansas City Star newspaper and referred to by Mencken in his letter of May 12, 12, 1947 appearing on page 2.

courtyard at the University of Kansas hospitals. The speaker reminded the audience "that it was Henry Mencken who first recognized the possibilities in Dr. Clendening's medical writings and persuaded Alfred Knopf, the publisher to invite the physician to write, 'The Human Body' ". (The circumstances and conditions of this encouragement are fleshed out earlier in this chapter)

(Mr. Haskell said, "It is appropriate that the memorial to Logan Clendening should take the form of a fountain. A fountain is no inert thing. It is full of life and interest, the movement of water suggests sparkle and animation- the qualities that were so conspicuous a part of Logan's endowment.)

The speaker observed that Dr. Clendening was man of quick perception who had the common touch as well as the "spark of genius". He reviewed his career, "having begun his practice in Kansas City in 1909 and four years late married Mss Dorothy Hixon, who shared his tastes and cultural interest. The great success of "The Human Body" led to his undertaking a syndicated newspaper column and consequently he felt obliged to give up the active practice of medicine for the work in which his genius was outstanding- writing, supplemented by teaching". Mr. Haskell was reported to have added that, " it was not that Dr. Clendening had failed as a competent practitioner, but that his personality better suited him for the direction in which his particular gifts led him." Dr. Haskell continued, "His newspaper column, 'Diet and Health' was appearing in 383 newspapers at the time of his death. I may confide to you that for the last two years, we at The Star have been looking for a medical column of the same quality to take its place and so far have not found one. Plenty of competently done medical columns are available but one would not think of reading them for pleasure."

(Mr. Haskell reminded his audience how Clendening's "literary qualities (appeared) in a discussion of theories that in the future we will satisfy our dietary needs with small pellets of condensed foods." Haskell quoted Clendening as answering, " with agreeable companions dining is certainly one of the most civilized of occupations. 'A good dinner and feasting reconciles everybody' quoth Mr. Pepys, and Dr. Johnson averred that "A man seldom thinks with more earnestness of anything than he does of his dinner.'" Mr. Haskell concluded that, " the important scientific contributions that he made in the field of his chosen profession were suffused with the joy of loving. It is this rare combination of qualities that those of us here today)most vividly remember." As Mr. Haskell finished his address, the water was turned on in the memorial fountain.)

Conclusion

How should the letter exchanges between H.L. Mencken and Logan Clendening be considered. An East coast editor/critic and a Midwestern doctor at first glance would have little in common. Here is represented a twenty one year running discourse, initially about a business venture to encourage the publication of a neophyte medical author on a popular medical topic, The Human Body. . He had previously written just one book , Methods of Treatment in 1924. This caught the eye of the editor and encouraged the doctor to continue filling the void this editor considered in the market of popular medical writing. Mencken, the editor, had encouraged many emerging writers such as Driser and F. Scott

Fitzgerald, by offering to publish their efforts in his magazine, The American Mercury. But in this case the relationship between the editor and doctor grew and blossomed into a mutual friendship for over two decades without interruption. For many of these emerging writers the relationship was brief and they set sail on their own course.. But the Clendening/Mencken relationship was warmed with the advancing years and sprouted new mutual interests . (While The Human Body, selling over one half million copies, was their first success together, Clendening went on to publish his own syndicated column in 383 newspapers.)

Clendening and Mencken went on to discuss many matters of mutual interest, the state of medicine, uses and abuses of alcohol and morphine. There was Clendening's letter of congratulation to Mencken on his upcoming marriage and his condolence letter four and a half years later on his wife's death.. Mencken's brief terse, typewritten letter to Clendening's wife Dorothy, eleven years later, on her husband's death is in striking contrast. On other subjects such as Leeuwenhoek's microscope, Clendening chided Mencken ever so gently on the facts of his review. On a lighter note is the exchanges between Dorothy and Mencken on there respective ponies, Dixie and Frank. And then there is Mencken's request for clarification of certain medically related terms, "to goose, Charley horse, and glass arm " for inclusion in his Supplement to the American Dictionary representing a fertile field for their exchanges. Finally there is their exchanges concerning their mutual health and the delightful injunction from Clendening to Mencken to take care of himself and " stop answering so many letters" (He is reputed to have written over 150,000 letters.)

As with other young authors that Mencken encouraged, his own irreverent, brisk, bombastic style may have rubbed off on them in their own writings and for Clendening this was no exception. There were however striking differences in their writing styles. Mencken's letters, always typewritten, impeccable in their spelling, absence of any typing miscues, focused and to the point, with usually an unrelated humorous irreverent political or other unrelated material acting as a final coda. This is to be contrasted with Clendening's more verbose, rambling style but preserving some Menkenese images,as well as his own characteristic writing.

Clendening prided himself on his knowledge of Mencken's literary out put, stating, " the fifth series of Prejudices were the best" and by frankly claiming "I believe that I am as familiar with your published works as anybody and I imagine I could pass an examination in them that you could not pass yourself."[23] This letter is an example. Some of this boasting may have been to flatter the more famous and older (by however only four years) colleague and get into his better graces but most of it was just honest appreciation of Mencken's style. Clendening often floated new ideas for writing articles to Mencken for his approval and evaluation. Finally, both of them are frequently entreating each other to get together. They seemed to enjoy each others letter writing exchanges and reports of their actual meeting together would be interesting.

The Clendening / Mencken letter exchanges represent only one example of the many more famous letter writing

---

[23] In a letter of March 31, 1943 to H.L. Mencken . "The Ruins of Carthage" referred to in this letter appeared in an Evening Sun's article by Mencken dated March 26, 1934 and later rewritten as "Vanishing Act" , chapter XVll in Heathen Days as part 3 in the Day's book trilogy , published in 1943.

exchanges in the literature. Examples such as Abigail Adams/ John Adams or Goethe/Charlotte von Stein come to mind. Reading the letters of others sharpens the underlying characterization of the respective letter writers and allow a view into private and hidden thoughts that are not available from other sources, almost the experience of a voyeur The art of letter writing, while not yet an entirely lost art, has lost much of its momentum and may be a dying species. This may be the unintended consequences of E-mail, blackberries and other more cryptic forms of communication of today. Such would be the loss of those who sift through archival material to surreptitiously gather the personal conversation of others for use in their own painting of a historical situation.

I would like to express my appreciation to Dr. Vincent Fitzpatrick, curator of the Mencken collection at the Enoch Pratt Library in Baltimore for making possible the review of the archives of Clendening/Mencken material and extensive advice and expert suggestions on the evaluation of this material. Permission for use of the archival material from the Mencken collection at the Pratt Library was obtained from Mrs. Averil Kadis, Rights and Permissions, The Estate of H. L. Mencken, Public Relations Division, Enoch Pratt Free Library, 400 Cathedral Street, Baltimore , MD 21201. Dawn McInnis, Rare Book Librarian of the Clendening History of Medicine Library for opportunity to review extensive material from their collection and providing photographs I also appreciated the cooperation and help of the Goucher College Library in Baltimore and the New York Public Library for the opportunity to evaluate and view their archival material.

Addendum

A reason I have included this essay on the echange of letters between Mencken and Clendening is that Mencken himself was a lover of classical music. Mencken was a lifelong amateur musician. In his preface to "Happy Days," he writes that he spent half a century boasting that he could read music by the age of 6. But, upon finding a bill for the family's first piano, he discovered that he didn't begin playing until he was 8. Still, he wrote music for much of his life. "When I think of anything properly describable as a beautiful idea, it is always in the form of music. H.L. Mencken, by his own humble admission, wasn't much of a piano player, but the Baltimore icon gave it his all, especially when his Saturday Night Club convened to make music and imbibe. He spent decades playing as regularly as possible with the Saturday Night Club, which was founded in 1904 and came to a halt sometime around 1950, six years before Mencken's death. Mencken's colleagues included fellow amateurs, as well as some pros, among them Gustav Strube, the first music director of the Baltimore Symphony; composer Louis Cheslock, a Peabody Conservatory faculty member; and Adolph Torovsky, band director of the Naval Academy club. For more than 40 years, club members regularly assembled to perform arrangements of the classics and pieces written by colleagues, creating in the process a legendary part of Baltimore's history. The Enoch Pratt Library holds in its archives 54 boxes of the club's music. The club played for instance the opening movement to Schubert's "Unfinished" Symphony -- scored for five strings, clarinet, flute, horn and piano. The faithful arrangements of the Schubert work and the opening of Beethoven's "Eroica" offered reminders of how serious Mencken's club members could be about their music. The group would play their favorite pieces and eventually compose a few of their own. Mencken's favorite piece was Beethoven's "Eroica," a line from which appears on the group's shield, which is both mysterious and funny and adorns one of the walls of the Mencken room at the Pratt. Although separated by thousands of miles from Baltimore to Kansas, a subject held in great endearment to both Mencken and Clendening was classical music which they wrote about in their exchange of letters.

Conclusion

By David B. Levy

The Silence in Music

Music is more than the volume of its notes, but rather an expression of the human soul. Longinus in his essay on the sublime (*perihupsos*) notes that the sublime is not a function of amplitude, but evokes the greatness of a higher power as when witnessing the starry heavens above or a raging sea. For philosophers like Kant (*Observations on the Feeling of the Beautiful and Sublime*, 1763) and Burke (*Enquiry into the Origin of Our Ideas of the Sublime and Beautiful*, 1757), beauty is aesthetically revealed in being conveyed to the senses by melodious music and various fine arts, whereby the pieces fit together like an intricate architectonic of a clock. We can understand more deeply the essential beauty of art via Plotinus' (*Enneads*) definition of the sublimity of beauty based on its ability to lift "minds" themselves above the realm of sense, to a higher (ethical) order. Music that is classified as "good" (not vulgar) according to the Rambam can do this. This is not merely aesthetics, but rather internalization of ethical principles into moral conduct or action, and striving to perfect virtuous character traits in thought, mind, emotion and deed, such as magnanimity, graciousness and empathy.

A profound definition of beauty is found in Plotinus' *Enneads* which articulates a beauty of a higher order represented in virtuous conduct of life in actions, character, pursuits of the intellect. The Greek word for a gentleman is *kalos kagathos*, meaning `beautifully' souled, from the word *kalos* (beauty), and vulgarity is indeed the lack of such beauty as a negative construct, *aperkalia*, meaning vulgarity. Thus Longinus in his work on the sublime (*peri-hupsos*) refers to the Hebrew Bible (Septuagint) that he read in the Alexandrian Library as the most "sublime" work ever, because its laws encourage the ennoblement of beautiful character that is informed by the sublimity of feelings and intellectual intimations conveyed by music. Thus it was thought by Plato that one mode in music produced the virtue of temperance in a person. Another mode in music lent itself to the courage of a warrior. The frigeon mode it was thought by Plato induced a person towards ascetic virtue. In Judaism such beauty of virtue can be expressed in acts such as passionate dance, sublime singing, and spiritual ecstasy etc all conveyed by the vehicle of music with therapeutic effect.

Objectives of Volume One

It is hoped that this volume

(1) Shows how Humanistic academic inquiry and historical study can inform and are informed by science and medicine

(2) Shows interplay between the gamut of Humanities (Music, Art, lit, etc) and practice of medicine

(3) Touches on Medical arts, Culture, and science as inter related disciplines whereby music, lit, history and philosophy give an important depth to scientific knowledge

(4) Offers a human context of medicine as an art not just a science

(5) Shows relevance of Humanities to make medicine more sensitive to the needs of others and how historical

research can widen consciousness of the broader picture of the evolution of medical discoveries

(6) Shows how an understanding and appreciation of the Humanities and historical research can enrich and deepen knowledge of the historical development of medicine and science

(7) Provides a recognition of the humanistic and cultural dimensions of the history of medicine and allied sciences

(8) Fosters a wider perspective of the importance of historical research as elucidated by the Medical Humanities

Medicine, Music and Healing

It is hoped that the reader of this first volume of a five volume work, will come to better appreciate the relationship between music and medicine, and thereby better understand that a physician, true to the medical arts, is both a healer of the body, as well as the soul as we say when in the *Shemonah Esreh* prayer or *Amidah* referring to G-d as the ultimate Healer beyond any human strategies or abilities:

> Heal us, O L-rd, and we will be healed; help us and we will be saved; for You are our praise. Grant complete cure and healing to all our wounds; for You, Almighty King, are a faithful and merciful healer. Blessed are You L-rd, who heals the sick of His people Israel.

As the OU website notes,[1] the eighth bracha recalls the words of the prophet Jeremiah. Jeremiah says, "Heal me, Hashem, and I will be healed; save me and I will be saved because You are my praise" (17:14). This verse is incorporated in its entirety into the prayer for health, only changing Jeremiah's singular to the plural ("heal us," "save us," etc.). *Refuah* (healing) refers to repairing the body; *yeshuah* (salvation) refers to reinvigorating us with strength and vigor as if new, in deliverance by the power that makes for redemption.

The intent of this verse appears to be that if God heals us, then we are truly healed for better, and God is the true Healer. No matter what the doctor may prescribe, we can only be healed if God wills it. If He heals us, we will be healed, otherwise not. If we are healed, He is the One to praise for doing so, not any hospital or pharmacy, although the physician is the conduit or messenger for bringing healing. (God expects us to do our part even though the ultimate result is up to Him.)

Rabbi Abromowitz writes:

> From context, Jeremiah was speaking of spiritual rather than physical health. Really, the two things are intertwined, as our physical well-being is inextricably ties to our spiritual selves...What we ask of God is to be healed in both body and soul.

As with teshuvah in the fifth blessing, we ask Hashem for a "refuah shelaimah" – *rufuat ha-guf ve-refuah ha-nefesh*, a complete recovery. I believe that a true recovery can be enhanced by our greater appreciation of and recognition of the therapeutic and intellectual components of music to

---

[1] Rabbi J. Abramowitz, www.ou.org/torah/tefillah/shemoneh-esrei/shemoneh/esrei_8, accessed February 11, 2018.

heal the mind and body. *"Refuah shelaimah"* is the wish that Jews universally grant the sick, and as such music can play an important role in this praxis.

Maimonides wrote a teshuvah on listening to music and believed that good music can heal particularly getting one like Saul out of melancholy. Combining healing of the body with spiritual healing of the soul thus can be assisted by music. And thus the Mishnah in משנה מסכת סנהדרין פרק ד states

וכל המקיים נפש אחת מישראל מעלה עליו הכתוב כאילו קיים עולם מלא ומפני שלום הבריות

The Mishnah attests that each person is a world themselves and that s/he who saves one life is as if they saved a whole world. This gives the imperative to heal a fundamental principle. If music speaks to higher worlds still, the worlds to which the soul can travel in spiritual ascent then so too music can play a role in saving us to envision a realm of beauty in this world and beyond that is so essential for the therapeutic healing process of any individual. Music no doubt, as the neurologist Oliver Sacks shows, effects the brain in more ways than mere "endorphins." Rather, good music is spiritual and can heal the body via a psycho-somatic understanding of health.

ABOUT THE AUTHOR

Robert I Levy MD

Physician

Affiliation: Johns Hopkins Medical School, Sinai Hospital, Osler Society, Society for the History of Medicine

Curriculum Vitae

Date Birth 3/18/1926 Sinai Hospital Baltimore MD

High School Baltimore City College 1944

Peabody Institute of Music Preparatory 1934-1942; Conservatory 1942-1944

US Navy June 1944-1946 USS Saratoga, Pharmacist 3rd class

Johns Hopkins University major Biology 1946-1949 Phi Beta Kappa

Johns Hopkins Medical School 1949-1953

Intern and Assistant Resident Barnes Hospital St. Louis 1953-1955

Fellow Dept of Pharmacology Johns Hopkins Medical School 1955-1957

Fellow Dept. of Nephrology Mass Memorial Hospital, Dr. Arnold Relman 1957-1958

Chief Resident in Medicine Sinai Hospital Baltimore MD
1958-1959

Performed the 2nd Hemodialysis in Baltimore MD

Private Practice Nephrology 1959-2002, performing renal
consultations in Baltimore Hospitals, Renal Biopsies,
Hemodialysis. In charge of Dialysis facility in Westminster
MD for 20 years. Member of Osler Society delivering many
papers over the years and at the American Association of
History of Medicine

Publications

Papers Written While in Medical School (1949 – 1953)

1. Dissimilation of glucose-1 phosphate and of fructose 1-6
   phosphate by isolate rat diaphragm and by cell free
   effluent from rat diaphragm: K.L. Zierler, R.I. Levy, and
   R. Andres.   Bulletin of the Johns Hopkins Hospital,
   82:7, 1953.

2. On the mechanism of action of alpha-tocopheryl
   phosphate with special reference to carbohydrate
   metabolism of striated muscle.

        a. Effect on the capacity of rat diaphragm to
   dissimilate hexose phosphate: K.L. Zierler, R.I. Levy,
   H.M. Anderson and J. L. Lillenthal. Bulletin of the Johns
   Hopkins Hospital, 92:32, 1953.
        b. Inhibition of insulin induced glycogenesis on
   isolated rat diaphragm. K. L, Zierler, R.I. Levy, J.L.
   Lillenthal. Bulletin of the Johns Hopkins Hospital, 92,
   41, 1953.

Papers Written during Nephrology Fellowship with Dr. Gilbert H. Mudge (1955-1957)

3. The effect of acid base balance on the diuresis produced by organic and inorganic mercurials : R.I. Levy, I. M. Weiner, and G. H. Mudge. Journal of Clinical Investigation, 37: 1016, 1958.

4. Studies on mercurial diuresis: renal excretion, acid stability and structure activity relationships of organic mercurials: I. M. Weiner, R.I. Levy, and G. H. Mudge. Journal of Pharmacology and Experimental Therapeutics, 138: 96, 1962

Papers Written During House Staff Training or Medical Practice (1959 – 2002)

5. Renal failure secondary to ethylene glycol poisoning: R.I. Levy. Journal of the American Medical Association, 173, 1210, 1960.

6. Steroid blocking agents as diuretic agents: R.I. Levy. Sinai Hospital Journal, 10:110, 1961.

7. Serum sodium concentration: Facts of Fancy: R.I. Levy. Indian Medical Journal, 1962, (October).

8. Lipids of the kidney, Blood and Urine in the Nephrotic Syndrome: R.I. Levy, Fifteenth Annual Conference on the Kidney 1964.

9. Antibiotics and Digitalis Administration in Uremia : R.I. Levy. Editorial – Maryland Medical Journal, 13, 1964.

10. Ethacrynic Acid in Pulmonary Edema: R.I. Levy, A.I.

Mendeloff, D. Turner. American Journal of Clinical Nutrition, 18: 20, 1966.

11. Studies in a Patient with Chyluria: R. I. Levy, A.I. Mendeloff, D. Turner. American Journal of Clinical Nutrition. 18:20, 1966.

12. Overwhelming Salicylate Intoxication in an Adult: R. I. Levy. Archives of Internal Medicine, 119, 1967.

13. Treatment of Hypercalcemia with Forced Saline Diuresis and Ethacrynic Acid. R.I. Levy, Proceeding of the American Society of Nephrology, 3rd Annual Meeding Washington, D. C. (Abstract) 0.40, 1969.

14. Clinical Spectrum of Lactic Acidosis. R.I. Levy, K. Dharmasena (Paper presented at Regional Meeting, American College of Physicians in Baltimore, MD, October, 1975.

15. Serum Chloride Analysis, Bromide Detection and the Diagnosis of Bromism: American Journal of Clinical Pathology, R.I. Levy, R.E. Wenk, Lustgarton, Pappas and Jackson. Vol. 65: 49, 1976

16. Ectopic ACTH, Prostatic Oat Cell Carcinoma and Marked Hypernatremia, R.E. Wenk, B.S. Bahagavan, R.I. Levy, D. Miller and W. Weisburger. Cancer, Vol. No 2, August , 1977.

17. Chyloperitoneum in a Peritoneal Dialysis Patient. American Journal of Kidney Diseases. Vol 38. No. 3 (Sept) 2001; pE 12.

Unpublished paper:

Mozart and Medicine at the End of the Eighteenth Century   R.I. Levy   1990

Presented at Sinai Hospital, Baltimore, MD

Some papers Written Following Retirement from Medical Practice, 2002

18. History of Sinai Hospital of Baltimore Maryland 1863 - 2009, Its Place in the History of Jewish Hospitals in America

19. William Osler's Mention of Basham's Mixture in the Treatment of Bright's Disease: Who was Basham and What was his Mixture

20 The Animal Chemists in the Circle of Richard Bright

21. Therapeutic Spectrum Available to Defining the Newly Recognized Clinical Entity: Bright's Disease

22. The Reception in Britain and on the Continent of Richard Bright's – Report of Medical Cases on Linking Dropsy, Coagulable Urine and Small Granular Kidneys as a Clinical Entity

23. A Garland of Ibids: the Use of Footnotes in the Medical Writings of Early Nineteenth Century Authors Who Established Bright's Disease a Clinical Entity

24. Sir William Osler: A Departure from His Reputation as a Therapeutic Conservative: The Treatment of Bright's Disease

25. Urinalysis as a Factor in the Establishment of the Clinical Entity of Bright's Disease in the Early 19th

Century

26. Pulvis Impecacunanhae et Opii – The Powder and the Buccaneer, Thomas Dover (1660-1742)

27. Sir William Osler's View of Pierre C.A. Louis' Recommendations for Bleeding in Pneumonia. The Paradox of Calling Louis's Method Iconoclastic, yet Continuing his own Practice of Bleeding in Some Cases of Pneumonia

28. The Doctor and the Newspaper Editor/Literary Critic: Correspondence between Logan Clendening, M.D. and H. L. Mencken, 1924 thru 1944 .

29. A conversation Between Two Leeches: Their History, Function and Contribution to Medical Therapy - Past and Present

30. Colour Indicators, Robert Boyle's Experimental History of Colours, Lignum Nephriticum (Presented at the Osler Society May 2010)

30. Nicholas Monardes, Guaiacum – The Holy Wood from the New World and the French Pox.

31. Chevalier John Taylor, J. S. Bach, George Frideric Handel -- Did the Chevalier Really Operate on both Bach and Handel for Cataracts with Disastrous Results ?

32. William A. Marburg's Contribution to Sir William's Osler's Love for books and Libraries

33. Dr. Theodore Billroth and Johannes Brahms: a musical friendship

34. William Harvey and his De Motu Cordis

35. Homer Smith and the Evolution of the kidney (presented at American Society of Urology 2017)

36. A tribute to Historian of Medicine Dr. John Conrad Hemmeter

Acknowledgments

I would like to dedicate the book to my parents, Dr. Robert I Levy Shlita and Ruth S. Levy (zl) who taught me to behave with *mentschlikeit*, refinement, and care for the needs of others, support my work in every way and placed value and utmost importance on critical thinking, scholarship, and quest in the life of the mind in search for *hokmah, binah, vedaat*. My parents taught me the importance of internalizing/sublimating into the *neshamah: derekh eretz, darkei noam* (pleasantness), respect for all (*kavod ha-briyot*), speaking with the purity (*loshon naki*), and striving towards *middot tovot* (good character traits). I am blessed to have wonderful parents whose magnanimity, graciousness, great souledness, *ahavas Hashem, ahavas Torah*, and *ahavas olam*, make their lives an example of kiddush Hashem.

I also wish to thank my eishet chayil Ariella Chasidah, from whose emunah and betachon I have much to learn, and whose practical knowledge (*phronesis*) serves as a balance to complement my more theoretical side: פיה פתחה בחכמה ותורת חסד על לשונה...אשה יראת ה' היא תתהלל.

I would also like to thank the publisher Mr. Sam Sapozhnik for having faith in this project and providing the opportunity to bring the ideas to light in the form of a publication. I also owe a debt of great gratitude to Dr. Hillel Abramson for his efforts in helping bring this multi-volume set to publication.

DBL

Cedarhurst, NY Adar 5778/March 2018, Purim

This book was set in Palatino, a typeface designed in 1949 by Herman Zapf. Reminiscent of Times New Roman, favored by scholarly publications, Palatino retains the elegance and *gravitas* of its predecessor yet exerts a subtle note of distinction and independence from the conventional.

www.ingramcontent.com/pod-product-compliance
Lightning Source LLC
Chambersburg PA
CBHW071258220526
45468CB00001B/188